Science

11+

Practice
Papers

The Publishers would like to thank the following for permission to reproduce copyright material.

Photo credits p15 © ATTILA KISBENEDEK/AFP/Getty Images **p20 (l)** © unclepodger/Thinkstock/iStockphoto/Getty Images **p20 (r)** © Jackie Barns-Graham **p21 (t)** © MIGUEL GARCIA SAAVED - Fotolia.com **p27** © Georgios Kollidas - Fotolia **p28 (b)** © TopFoto **p37 (b)** © Denis Doyle/Getty Image/Thinkstock **p46 (b)** © Jackie Barns-Graham

Every effort has been made to trace all copyright holders, but if any have been inadvertently overlooked, the Publishers will be pleased to make the necessary arrangements at the first opportunity.

Although every effort has been made to ensure that website addresses are correct at time of going to press, Galore Park cannot be held responsible for the content of any website mentioned in this book. It is sometimes possible to find a relocated web page by typing in the address of the home page for a website in the URL window of your browser.

Hachette UK's policy is to use papers that are natural, renewable and recyclable products and made from wood grown in sustainable forests. The logging and manufacturing processes are expected to conform to the environmental regulations of the country of origin.

Orders: please contact Hachette UK Distribution, Hely Hutchinson Centre, Milton Road, Didcot, Oxfordshire, OX11 7HH. Telephone: +44 (0)1235 827827. Email education@hachette.co.uk. Lines are open from 9 a.m. to 5 p.m., Monday to Friday. You can also order through our website: www.hoddereducation.com

ISBN: 978 1 4718 4928 2

© Jackie Barns-Graham 2016
Published by Galore Park Publishing Ltd,
An Hachette UK Company
Carmelite House
50 Victoria Embankment
London EC4Y 0DZ
www.galorepark.co.uk
Impression number   10 9 8 7
Year   2022

All rights reserved. Apart from any use permitted under UK copyright law, no part of this publication may be reproduced or transmitted in any form or by any means, electronic or mechanical, including photocopying and recording, or held within any information storage and retrieval system, without permission in writing from the publisher or under licence from the Copyright Licensing Agency Limited. Further details of such licences (for reprographic reproduction) may be obtained from the Copyright Licensing Agency Limited, www.cla.co.uk

Illustrations by Integra Software Services Ltd
Typeset in India
Printed and bound by CPI Group (UK) Ltd, Croydon, CR0 4YY

A catalogue record for this title is available from the British Library.

Name: _____

11+

Science

Practice Papers

Jackie Barns-Graham

GALORE
PARK

AN HACHETTE UK COMPANY

Contents and progress record

Difficulty	Score	Time	Notes
easy	☐ / 30	☐ : ☐	
more difficult	☐ / 50	☐ : ☐	
easy	☐ / 30	☐ : ☐	
more difficult	☐ / 50	☐ : ☐	
easy	☐ / 30	☐ : ☐	
more difficult	☐ / 50	☐ : ☐	
11+ Practice Paper	☐ / 80	☐ : ☐	
11+ Practice Paper	☐ / 80	☐ : ☐	
11+ Practice Paper	☐ / 80	☐ : ☐	
11+ Practice Paper	☐ / 80	☐ : ☐	
11+ Practice Paper	☐ / 80	☐ : ☐	
11+ Mock Exam	☐ / 80	☐ : ☐	

How to use this book

Introduction

These *Practice Papers* have been written to provide final preparation for your 11+ Science test.

This book includes six single-subject training tests for Biology, Chemistry and Physics (two for each subject) and five 11+ papers modelled on the actual exam papers. The single-subject training tests are divided into one simpler paper which tests your basic understanding of the subject and a second, more challenging paper, which assesses your ability to analyse and apply your scientific knowledge.

The *Practice Papers* will help you to:

- become familiar with the way 11+ tests are presented
- build your confidence in answering the variety of questions set
- work with increasingly difficult questions
- tackle questions presented in different ways
- build up your speed in answering questions to the timing expected in the 11+ tests.

The final test in the book is designed to be a Mock Exam to test your preparation shortly before you sit the real paper for entrance to your next school. As in real papers, it contains questions that are similar to those you have already practised using this book. This should be your experience when you sit your real 11+ paper.

Pre-Test and the 11+ entrance exams

The Galore Park 11+ series is designed for Pre-Tests and 11+ entrance exams for admission into independent schools. These exams are often the same as those set by local grammar schools. Many schools set the ISEB 11+ Science exam, although some schools set their own and it is possible that if you are applying for more than one school, you will encounter more than one of type of test.

To give you the best chance of success in these assessments, Galore Park has worked with 11+ tutors, independent school teachers, test writers and specialist authors to create these *Practice Papers*. The content covers the National Curriculum Programmes of Study for Key Stages 1 and 2 up to the end of Year 5, as well as the ISEB syllabus to the end of Year 5 and the 11+ examinable material in Year 6.

For parents

These *Practice Papers* have been written to help both you and your child prepare for both Pre-Test and 11+ entrance exams.

For your child to get maximum benefit from these tests, they should complete them in conditions as close as possible to those they will face in the exams, as described in the 'Working through the book' section on the next page.

The time allowed to complete questions gets shorter as the book progresses to build up speed and confidence.

Some of these timings are very demanding and reviewing the tests again after completing the book (even though your child will have some familiarity with the questions) can be helpful, to demonstrate how their speed has improved through practice.

For teachers and tutors

This book has been written for teachers and tutors working with children preparing for both Pre-Tests and 11+ entrance exams. The syllabus coverage to the end of Year 5 (both National Curriculum and ISEB) and ISEB Year 6 examinable material, has been compressively reviewed to ensure questions are asked on all subject areas.

Working through the book

The **contents and progress record** helps you to understand the purpose of each test and track your progress. Always read the information in this chart before beginning a test as this will give you an idea of the content and how challenging the test will be.

You may find some of the questions hard, but don't worry – these tests are designed to build up your skills and speed. Agree with your parents on a good time to take the test and set a timer. Prepare for each test as if you are actually going to sit your 11+ (see 'Test day tips' below):

- Complete the test with a timer, in a quiet room, noting down how long it takes you, writing your answers in pencil. Even though timings are given, you should complete ALL the questions.
- Mark the test using the answers at the back of the book.
- Go through the test again with a friend or parent and talk about the difficult questions.
- Have another go at the questions you found difficult and read the answers carefully to find out what to look for next time.

The **answers** are designed to be cut out so that you can mark your papers easily. Do not look at the answers until you have attempted a whole paper. Each answer has a full explanation so you can understand why you might have answered incorrectly.

When you have finished a test, turn back to the contents and progress record and fill in the boxes:

- Write your total number of marks in the 'Score' box.
- Note the time you took to complete ALL the questions in the 'Time' box.

After completing the book you may want to go back to the earlier papers and have another go to see how much you have improved!

Test day tips

Take time to prepare yourself the day before you go for the test. Remember to take sharpened pencils, an eraser, a calculator and a watch to time yourself (if you are allowed – there is usually a clock present in the exam room in most schools). Take a bottle of water in with you, if this is allowed, as this will help to keep you hydrated and improve concentration levels.

... and don't forget to have breakfast before you go!

Continue your learning journey

When you've completed these *Practice Papers*, you can carry on your learning right up until exam day with the following resources.

The *Revision Guide* covers all the topics you will need to know in your 11+ Science exam. The book is an essential tool to help you revise the topics you have been taught in school, plus some extra skills needed for the ISEB exams. The examiner's tips in the book will give you ideas for how to stand head and shoulders above the other candidates.

● Paper 1: Biology

Test time: 30:00

1   Underline the option that best completes each of the following sentences.

(a) A substance required by a plant to make its own food is

  a   carbon dioxide   b   fertiliser   c   oxygen   d   soil

(b) In a food chain the plant is always a

  a   consumer   b   herbivore   c   omnivore   d   producer

(c) A badger is described as nocturnal because it

  a   hibernates when it is cold   b   hunts for food during the day

  c   hunts for food during the night   d   sleeps during the night

(d) An example of a vertebrate animal is a

  a   bee   b   snail   c   snake   d   worm

(e) Some flowers have long, dangling stamen to encourage

  a   fertilisation   b   germination   c   pollination   d   seed dispersal

(f) Lack of vitamin C in the diet causes

  a   blindness   b   rickets   c   scurvy   d   spots   (6)

2   The diagram below shows a cross-section of a flower.

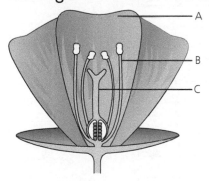

(a) Complete the table below with the names of each of the parts A, B and C and describe what each part does.

	Name of part	What the part does
A		
B		
C		

(3)

Turn over to the next page.

(b) Fill in the blanks to complete the sentence below.

The cells of the leaf and stem contain a green pigment called

_____ , which absorbs light energy from the

_____ in a process called _____ . (3)

3 This question is about digestion and food.

(a) Using straight lines, match the boxes with the names of parts of the digestive system to the boxes stating what each part does.

Stomach	This connects mouth to stomach
Small intestine	Where water is absorbed and faeces are formed
Mouth	Where food is chewed and saliva added
Esophagus	These tear, cut and grind food
Large intestine	Where food is absorbed into the bloodstream
Teeth	Where food is churned and broken down

(3)

Fibre is an important part of a healthy diet.

(b) (i) Name one food that is a good source of fibre.

_____ (1)

(ii) Explain why fibre is important in the diet.

_____

_____ (2)

4 Choose words or phrases from the box below to complete the following questions. Each answer may be used once, more than once or not at all.

adaptation	asexual	caterpillar	endangered
extinct	hibernation	larva	maggot
migration	Moon	pupa	sexual
Sun	variation		

(a) In a food chain the energy originally comes from the _____. (1)

(b) Breeding programmes in zoos are important to conserve _____ species. (1)

(c) When an insect hatches the emerging young form is usually called a

_____. (1)

(d) Using a cutting to grow a new plant is an example of _____ reproduction. (1)

(e) A tortoise becoming dormant in the winter is an example of

_____. (1)

(f) A camel having large feet to stop it sinking into the sand is an example of

_____. (1)

5 Look at the food chain and then answer the questions on the next page.

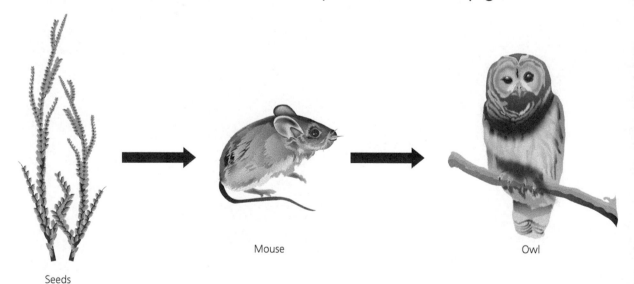

Seeds   Mouse   Owl

Turn over to the next page.

Using all of the words in the box, describe fully the feeding relationships of each of the organisms in the food chain shown in the diagram on the previous page.

carnivore	consumer	herbivore	predator	prey	producer

(a) (i) The plant is a ____________________ (1)

(ii) The mouse is a ____________________ (2)

____

(iii) The owl is a ____________________ (2)

____

(b) State what the arrows represent in the food chain.

____________________ (1)

Record your results and move on to the next paper.

Score ☐ / 30 Time ☐ : ☐

Paper 2: Biology

Test time: 50:00

1 Ann and Sophie had been told that the ivy growing around their house had different sized leaves when it was growing in full sunlight compared with when it was growing in the shade.

(a) Which two raw materials are needed by the leaf for plant growth?

_____

_____ (2)

(b) What is the name of the green pigment found in the green parts of the plant?

_____ (1)

(c) What is the role of the green pigment in the leaf?

_____ (1)

Sophie predicted that the ivy in full sunlight would have bigger leaves.
Ann predicted that the ivy would have bigger leaves in the shade.

(d) Suggest who could be correct. Explain your answer.

_____

_____ (2)

Turn over to the next page.

They decided to test their predictions.

They collected 50 leaves growing in the light and 50 leaves growing in the dark.

They measured the length and width of each leaf using a ruler.

They then found the mean (average) of their results.

length, in mm

width, in mm

(e) Why did they use the mean results of 50 leaves rather than just comparing two leaves?

_____

_____ (2)

Their results are shown below.

Leaves in the light		Leaves in the shade	
mean length, in mm	mean width, in mm	mean length, in mm	mean width, in mm
63	56	42	38

70
60
50
40
30
20
10
0

| Mean length, in mm | Mean width, in mm | Mean length, in mm | Mean width, in mm |
| Leaves in the light | | Leaves in the shade | |

Complete the bar chart above to represent their results.

(f) (i) Label the vertical axis. (1)

   (ii) Complete the bar chart to represent the data in the table. (2)

In deserts, grazing animals try to obtain water and food by eating plants that are adapted to store water in their stems. Many cacti have fleshy stems adapted to store water, while the leaves are small and adapted to protect the plant from grazing animals.

(g) Label a leaf on the picture of the cactus below. (1)

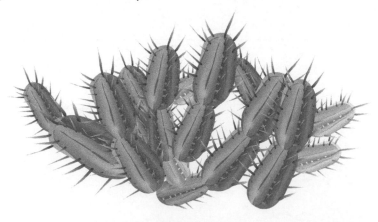

2 Read the passage about the animal behaviourist, Jane Goodall, and then answer the questions.

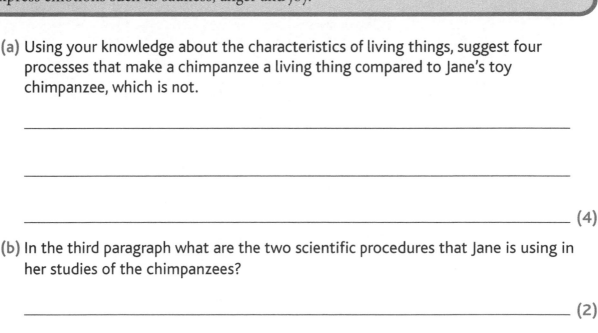

When Jane Goodall was a child she dreamt of going to Africa to see some of her favourite animals in the wild. She particularly liked chimpanzees. One of her favourite toys as a child was a toy chimpanzee, which she loved to play with.

When she grew up she went to Africa and met the British archaeologist Louis Leakey who offered her a job studying chimpanzees.

Jane had no formal training or education. This may have actually helped her as she had her own unique way of observing and recording the chimps' actions and behaviours. Jane spent the next 40 years of her life studying chimpanzees. She discovered many new and interesting things about the animals.

Jane observed a chimp putting grass into a termite hole in order to catch termites to eat. She also saw chimps remove leaves from twigs in order to make a tool. This was the first time that animals had been observed using and making tools.

Jane also discovered that chimpanzees hunted for meat. They would hunt as packs, trap animals, and then kill them for food. It had been thought before this that chimps only ate plants.

Jane observed many different personalities in the chimpanzee community. Some were kind, quiet and generous, while others were bullies and aggressive. She saw the chimps express emotions such as sadness, anger and joy.

(a) Using your knowledge about the characteristics of living things, suggest four processes that make a chimpanzee a living thing compared to Jane's toy chimpanzee, which is not.

_____

_____

_____ (4)

(b) In the third paragraph what are the two scientific procedures that Jane is using in her studies of the chimpanzees?

_____ (2)

Turn over to the next page.

(c) Summarise below Jane's three main discoveries about the behaviour of chimpanzees:

1. _____

_____ (1)

2. _____

_____ (1)

3. _____

_____ (1)

(d) Suggest how one of the behaviours in your answers in part (c) has contributed to the survival of chimpanzees in the wild.

_____

_____ (2)

3 (a) Use the key below to identify the seeds from plants A to G.
Write your answers in the table.

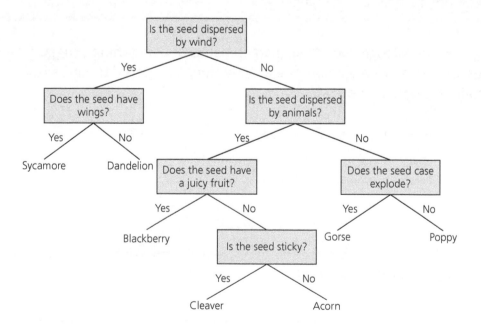

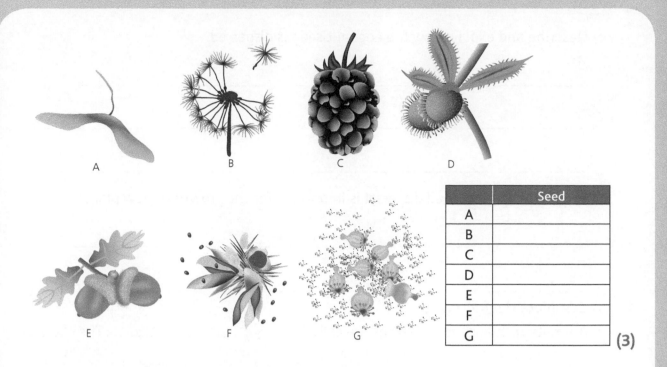

	Seed
A	
B	
C	
D	
E	
F	
G	

(3)

Look at the food chain below.

acorn ⟶ squirrel ⟶ hawk

(b) Using information from the food chain and your own knowledge, explain how acorns can be dispersed by squirrels.

_____

_____

_____ (3)

The diagram below is of a germinating coconut.

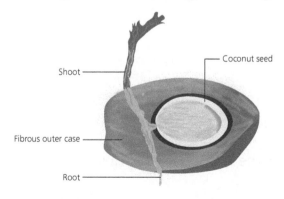

The part you can eat is the seed and it is in the middle of the diagram.

You do not often see the outer case when you buy a coconut but the fibres are often used to make doormats.

The fibrous outer case floats in water.

Turn over to the next page.

(c) Describe and explain how the coconut seed is dispersed.

_____

_____

_____ (3)

(d) Suggest why good seed dispersal is important for the growth of new plants.

_____

_____ (2)

4 (a) Complete the following:

Animals with internal skeletons are called _____. The skeleton

is important in providing _____, _____

and _____. (2)

(b) On the diagram of the human skeleton below label the skull, backbone, ribcage, pelvis, collarbone and shoulder blade.

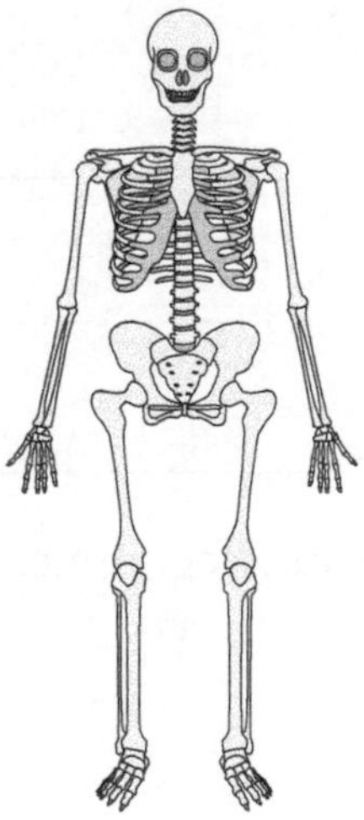

(3)

A caterpillar does not have a skeleton.
The diagram below shows a caterpillar moving.

(c) Label the diagram to describe fully how the caterpillar is moving. (3)

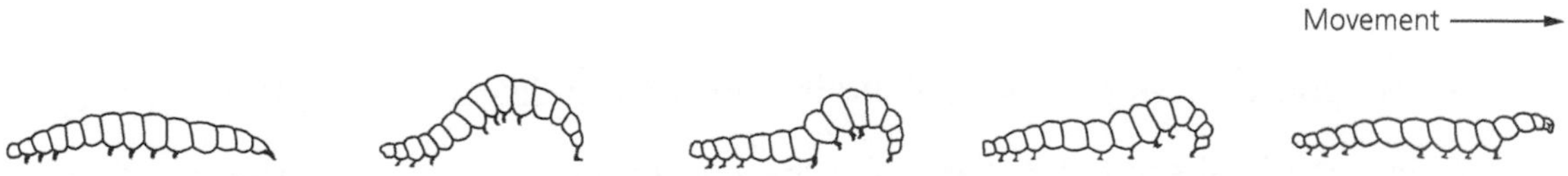

Movement ⟶

5 Humans can have both positive (+) and negative (−) effects on the environment.
Complete the table below to show in the second column whether the human action is positive or negative and then explain the consequences of the human action in the third column.

The first row has been done for you.

Human action	+ or − effect	Consequence for the environment
putting up bird boxes	+	provides places for birds to nest in areas where their normal habitat has been destroyed
keeping endangered species in zoos		
cutting down forests		
burning fossil fuels		
creating National Parks		

(8)

Record your results and move on to the next paper.

Score ☐ / 50 Time ☐ : ☐

Paper 3: Chemistry

Test time: 30:00

1 Underline the option that best completes each of the following sentences.

(a) Soil particles of different sizes can be separated by

    a decanting   b dissolving   c filtration   d sieving

(b) When water changes into a gas it is

    a boiling   b condensing   c evaporating   d melting

(c) In syrup, a mixture of sugar dissolved in water, the water is the

    a solid   b solute   c solution   d solvent

(d) An example of a solid fossil fuel is

    a coal   b crude oil   c natural gas   d wood

(e) An example of a man-made material is

    a cotton   b leather   c nylon   d rubber

(f) The body temperature of a healthy human is

    a 30–31°C   b 33–35°C   c 36–37°C   d 38–39°C (6)

2 The pictures below show two different rocks A and B.

A

B

(a) Complete the table below.

	Does it have grains or crystals?	Is the rock igneous or sedimentary?
A		
B		

(4)

(b) Underline the word, **in bold**, that makes a correct sentence about fossils, in each of the statements below.

(i) Fossils are formed when things that have lived are trapped in **igneous/ sedimentary** rock.

(ii) Usually only the **hard/soft** parts of organisms are preserved. **(2)**

3 This question is about the properties of materials.

(a) Using straight lines, match the boxes to make four sentences.

Granite can be used for kitchen work surfaces because	it is strong.
Steel is used for building bridges because	it is flexible.
Wood or plastic can be used for saucepan handles because	it is hard and does not scratch.
Cotton thread is used for sewing because	it is a good insulator of heat.

(2)

(b) Suggest two reasons why glass is a suitable material for the window in the picture above.

_____

_____ **(2)**

Turn over to the next page.

(c) A boy misbehaving tried to scratch his name on the glass. He used a pencil but the mark rubbed off. It worked when he used the metal that held an eraser to the other end of the pencil.

Complete the following statements about the hardness of glass, pencil lead and metal.

The glass is _____ than the pencil lead.

The pencil lead is _____ than the metal. (2)

4 For each of the examples below state whether a physical or chemical change is being described.

(a) Chocolate melting in your mouth. _____ (1)

(b) Making concrete from cement, sand and gravel. _____ (1)

(c) The flame on a candle. _____ (1)

(d) The wax melting when a candle burns. _____ (1)

(e) A strawberry turning from green to red when it ripens.

_____ (1)

(f) Paint powder mixed with water in art lessons. _____ (1)

5 Look at the picture below showing how heat is lost from a typical house. The sizes of the arrows show how much heat is being lost.

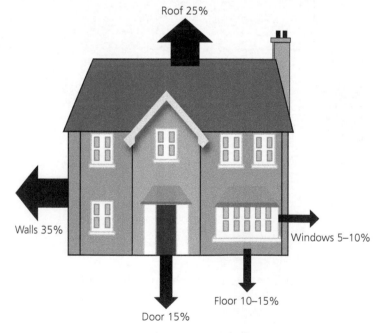

Roof 25%

Walls 35%

Windows 5–10%

Door 15%

Floor 10–15%

(a) Complete the table below with suggestions about the different methods you could use to reduce the heat loss from the house.
One has been done for you.

Part of house	Suggestions
roof	
walls	
windows	
door	
floor	thick carpets on floorboards with a space/ cavity below

(4)

(b) Write a brief sentence to summarise how your suggestions all work to reduce heat loss from the house.

_____

_____ (2)

Record your results and move on to the next paper.

Score ☐ / 30 Time ☐ : ☐

Paper 4: Chemistry

1 Andrew and Beth's teacher complained that the laces on their shoes were always coming undone.

They discovered that they were using two different types of laces.

Flat cotton laces

Round cotton laces

(a) Suggest two properties needed by a material used to make shoelaces.

_____ (2)

They decided to investigate to see which shoelace was better.

To start they made a prediction about which type of lace would last the longest before coming undone.

(b) (i) Suggest a prediction they could have made.

_____ (1)

(ii) Explain why you have suggested this prediction.

_____ (1)

Their laces seemed to always come undone as they walked to school so they decided to see how long their laces stayed tied each morning.

(c) (i) In their investigation the:
         independent variable is _____

         dependent variable is _____ (2)

    (ii) Describe how they should make their experiment a fair test.

         _____

         _____

         _____

         _____ (3)

    (iii) Suggest how they could make their results more reliable.

         _____ (2)

(d) Suggest a further investigation they could do to explore the problem of undone shoelaces.

    _____ (1)

2 Asha noticed gritty bits in the bath water when she added her orange bath salts.

She thought that some of the bath salt was not dissolving.

She decided to test the solubility of the bath salts at different temperatures.

She added the bath salt to a 1 litre jug of water until no more would dissolve, changing the temperature of the water for each test.

Turn over to the next page.

Here are her results:

Temperature, in °C	Amount dissolved, in g
20	20
30	30
40	40
50	50
60	55

(a) On the grid below:
    (i) plot the points from the table   (2)
    (ii) join your points with a suitable line.   (2)

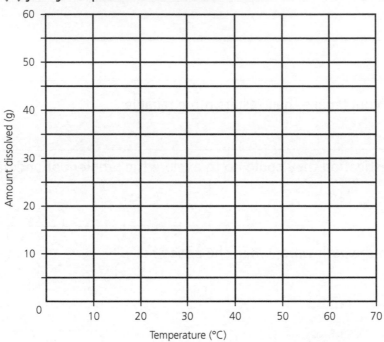

(b) How much bath salt would you expect to dissolve:

    (i) at 45 °C? _____ g   (1)

    (ii) at 70 °C? _____ g   (1)
    Asha measured the temperature of her bath water and it was 37 °C.

(c) Suggest why this is a comfortable temperature for a bath.

_____

_____ (2)

    The volume of the water in her bath is 80 l.

(d) If she adds 50 g of bath salts to her bath, will it all dissolve? _____
    Explain your answer.

_____

_____ (2)

(e) Complete the following:

The gritty bits in Asha's bath are _____ solids and could be

removed from the mixture in Asha's jug by _____. (2)

3 Read the following text about Antoine Lavoisier and then answer the questions.

> Antoine Lavoisier was born in Paris on August 26, 1743.
>
> In 1775, Lavoisier set up a laboratory where he used experiments and precise measurements to discover new facts in science.
>
> He demonstrated that there was a material called oxygen that plays a major role in burning.
>
> He discovered that we breathe in oxygen and breathe out carbon dioxide.
>
> He discovered that water is made of two materials: hydrogen and oxygen.

(a) How did Lavoisier discover new facts in science?

_____

_____ (2)

Combustion is another name for burning.

(b) Which material did Lavoisier discover that plays a major role in combustion?

_____ (1)

Lavoisier also showed that the mass of materials before a physical change is equal to the mass of the materials after the change.

(c) Calculate the mass of a solution where 15 g of salt has been added to 150 cm³ of water. (1 cm³ of water weighs 1 g)

_____ (1)

(d) What materials is water made from? _____ (2)

Many of Lavoisier's experiments involved the burning of different materials.

Complete the following sentences about burning by underlining the best options shown in bold.

(e) Burning is an example of **reversible/non-reversible** change. A material burned to produce heat is known as a **fuel/fire**. A material produced both in burning and in breathing is **carbon dioxide/oxygen**. (3)

Turn over to the next page.

4 Look at the diagram of the thermometer.

(a) (i) What is the temperature shown on the thermometer? _____   (1)
     (ii) Draw an arrow to show the temperature, −7 °C, on the thermometer.   (1)
     Pure water freezes at 0 °C. Water expands on freezing.

(b) Suggest two possible consequences of water expanding at it freezes.

_____

_____ (2)

   Salt is often put on the roads in winter to melt ice.

   The salt lowers the freezing point of water to about −9 °C.

(c) Once the salt is added, will the roads be icy when it is −7 °C? _____   (1)
   The winter in 1963 was the coldest in the UK for over 200 years, with temperatures
   falling to −20 °C.

   Look at the picture below showing ice forming on the sea off the Kent coast in 1963.

(d) Explain, using data from the question and your own knowledge, why the sea froze in 1963.

_____

_____

_____ (3)

Look at the table below showing information about four other substances that can lower the freezing point of water:

Substance	Works down to:	Advantages	Disadvantages
A	−7C	fertiliser	damages concrete
B	−29°C	melts ice faster than salt	attracts moisture, makes surfaces slippery below −18°C (0°F)
C	−9°C	safest for concrete and vegetation	works better to prevent re-icing than as ice remover
D	−15°C	melts ice faster than ordinary salt (sodium chloride)	attracts moisture

(e) (i) Which substances could be used to melt ice on the roads in a winter when temperatures fall to −12°C?

_____ (2)

(ii) Which substance should be avoided on concrete driveways?

_____ (1)

Explain your answer.

_____ (1)

Turn over to the next page.

5   Complete the diagram to show:
    (a) the arrangement of the particles in the two blank boxes   (2)
    (b) the names of the changes of state as indicated by the arrows.   (3)

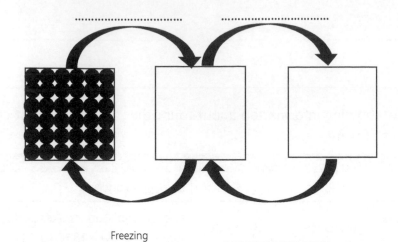

Freezing

● **Paper 5: Physics**

Test time: 30:00

1 Underline the option that best completes each of the following sentences.

   (a) An object that is always non-luminous is

   a   a desk lamp   b   a mirror

   c   a television   d   the Sun

   (b) The diagram below shows the wiring in an electric plug.

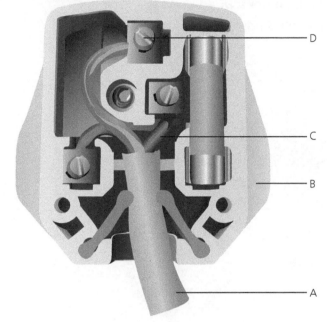

   A material in the plug that conducts electricity is

   A   B   C   D

   (c) The upward force acting upon a floating ship is

   a   upforce   b   uplift

   c   upthrust   d   water resistance

   (d) The force diagram that best represents the pulley lifting its load is

   a 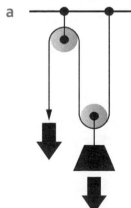 b  c 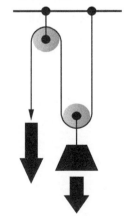 d

Turn over to the next page.

(e) Faster vibrations produce a sound that is
  a higher pitched
  c lower pitched
  b louder
  d quieter

(f) The Milky Way is a
  a comet
  c solar system
  b galaxy
  d star   (6)

2 When a cyclist rides a bicycle friction can act in a number of ways:
  A between the wheels and the road
  B between the air and the cyclist
  C between the wheel rim and the brake blocks when braking
  D between the moving parts of the chain and cogs turning the wheels.

Complete the table below for the situations B, C and D. Tick the appropriate box in the first two columns and then explain your answer in the last column.

A has been done for you.

	Useful friction	Nuisance friction	Explanation
A	✓		Friction between the wheels and the road is called grip, it prevents the bike skidding and allows the cyclist to control the motion of the bike.
B			
C			
D			

(6)

3 (a) Underline the word, **in bold**, that makes a correct sentence about magnets, in each of the following:
      (i) The north-seeking pole of a magnet will **attract/repel** another north-seeking pole.
      (ii) The south-seeking pole of a magnet will **attract/repel** an iron bar.   (2)
   (b) Draw lines pointing to the poles on each of the magnets.

(3)

   Magnetic forces act at a distance.

   (c) Give an everyday example of where magnetic forces act at a distance.

   _____ (1)

4 This question is about the Earth and space.

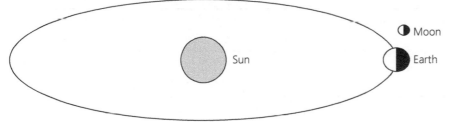

   (a) The roughly circular path that the Earth takes around the Sun is known as its

   _____.   (1)

   (b) Describe how day and night occur on Earth.

   _____

   _____ (2)

   (c) It takes approximately _____ for the Moon to go around the Earth.   (1)

   (d) State two observations that are evidence that the Earth is spherical.

   _____

   _____ (2)

Turn over to the next page.

5 (a) In the space below draw a circuit diagram with two cells, an open switch and a buzzer connected in series. (4)

Although everything in the circuit was connected correctly, when the switch was closed the buzzer did not work.

(b) Suggest two reasons why the buzzer did not work.

____

____

____

____ (2)

Record your results and move on to the next paper.

Score [ ] / 30 Time [ ] : [ ]

Paper 6: Physics

1   Hannah and Maya were testing to see what happened to the brightness of a bulb when different materials were connected in an electrical circuit.
They set up a circuit like the one below.

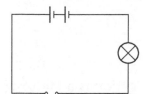

They connected different materials in the gap in the circuit.
Electrical conductors are materials that conduct electricity and the bulb will light.

(a) What do we call materials that do not conduct electricity?

_____ (1)

They observed the following:

Material	Bulb
steel paper clip	bright
2p coin	bright
eraser	off
green colouring pencil	off
2B drawing pencil	dim

(b) (i) What conclusion can they make about the metal materials in their circuit?

_____ (1)

The pencils were connected as shown below.

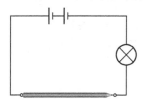

(ii) What conclusions can they make about each of the two pencils in their circuit?

_____

_____ (2)

(iii) What overall conclusion can they make about all of the materials that conducted electricity?

_____ (1)

Turn over to the next page.

Hannah and Maya decided to investigate further to see if all metals were good conductors of electricity.

Their teacher gave them three pieces of wire, A, B and C, made of different metals.

The table below shows the brightness of the bulb when each piece of wire was connected in their circuit.

Wire	Bulb
A	bright
B	normal
C	very dim

(c) (i) How should Hannah and Maya make sure that their investigation is a fair test?

_____

_____ (3)

(ii) What conclusion can they make about the metal wires in their circuit?

_____ (1)

The metal in wire A is used in household wiring.
(iii) What metal is wire A most likely to be?

_____ (1)

2   We see luminous objects when light from them enters our eyes.
Non-luminous objects are seen by reflected light.
The box below contains a selection of objects.

black tinted window	computer screensaver	mirror
radio	television picture	the Sun

(a) Decide how each object is seen and write your answers in the table.

Seen as a luminous source	Seen by reflected light

(3)

(b) Add a light ray to the diagram below to show how the girl is seeing the lamp. (2)

Look at the picture below.

(c) Why are we advised not to look directly at the Sun during an eclipse?

_____ (1)

Turn over to the next page.

Look at this diagram showing a torch, a shadow puppet and a screen.

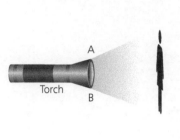

Screen

(d) (i) Draw two light rays to show the height of the shadow that will be seen on the screen. (1)

(ii) Sketch another similar shadow puppet closer to the screen and draw two more light rays to show the height of the shadow now. (1)

(iii) What can you conclude from the sizes of the two shadows?

_____ (2)

3 This question is about sound and hearing.

(a) Complete the following paragraph using words from the box.
Words may be used once, more than once or not at all.

| higher   louder   lower   pitch   quieter   volume |

Sounds are made when objects vibrate. Increasing the size of vibration makes

the sound _____. Increasing the speed of vibration

makes the sound _____ pitched.

When playing a guitar, shortening the string increases the _____

and plucking it gently will make a _____ sound. A thick string

has a _____ pitched sound than a thin string. Tightening the

string will make the note _____. (6)

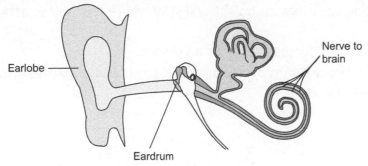

Look at the diagram of the ear.

(b) (i) Suggest a function of the earlobe.

_____ (1)

(ii) Which part of the ear vibrates first when a sound is heard?

_____ (1)

4   Sir Isaac Newton was an important scientist.
He is famous for having discovered gravity and the laws of motion.
He is one of the few scientists to have a unit named after him.

(a) What is the newton a unit of? _____   (1)

Newton is probably most famous for discovering gravity.

(b) What do you understand by the word 'gravity'?

_____

_____ (2)

(c) Complete the following passage about gravity.

Gravitational _____ keeps the Moon in _____

around the _____ and the _____ on their

paths around the _____.   (5)

Turn over to the next page.

Newton's first law of motion says that an object will stay where it is unless acted on by a force.

Arrows are used to represent the forces acting on an object.

(d) Add four labelled arrows to the diagram of a submarine, to show the names and directions of the forces acting upon it.

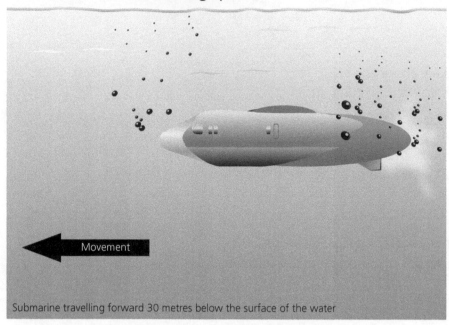

Movement

Submarine travelling forward 30 metres below the surface of the water

(4)

5 The picture below shows a sledge being pulled by dogs.
   A force meter was added to the harness to show the force needed to start moving the sledge.

The force needed to start moving different loads on the sledge was measured.
The results are shown below.

Load on sledge, in N	Force needed to start moving the sledge, in N
0	100
250	130
500	150
750	180
1000	200
1250	230

(a) On the grid below:
  (i) add scales to each axis (2)
  (ii) plot the points from the table (3)
  (iii) draw a straight line that best fits the points. (1)

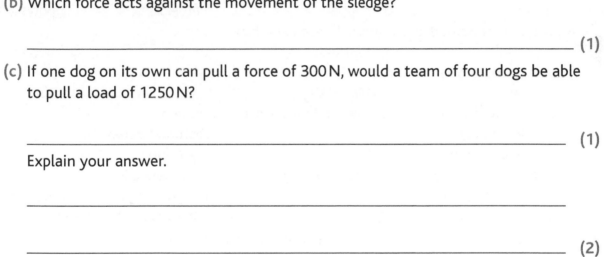

Force needed to move sledge (N)

Load (N)

(b) Which force acts against the movement of the sledge?

_____ (1)

(c) If one dog on its own can pull a force of 300 N, would a team of four dogs be able to pull a load of 1250 N?

_____ (1)

Explain your answer.

_____

_____ (2)

Record your results and move on to the next paper.

Score [   ] / 50 Time [   ] : [   ]

Paper 7: 11+ Practice Paper

Test time: 60:00

1   Underline the option that best completes each of the following sentences.

(a) Fish is a good source of

    a   carbohydate   b   fibre   c   minerals   d   protein

(b) For a seed to germinate it requires

    a   air, water and darkness   b   air, water and warmth
    c   water, light and warmth   d   water, warmth and darkness

(c) An object that lets only some light pass through it is described as

    a   luminous   b   opaque   c   translucent   d   transparent

(d) Sound cannot travel through

    a   a vacuum   b   air   c   metal   d   wood

(e) In seawater, salt is the

    a   solid   b   solute   c   solution   d   solvent

(f) The material that is not a good thermal insulator is

    a   air   b   copper   c   feathers   d   polystyrene

(g) A dog is classified as a mammal because it has

    a   a tail   b   four legs   c   fur   d   two ears

(h) An example of a star is the

    a   Earth   b   Milky Way   c   Moon   d   Sun

(i) A swallow flying north to breed in Europe is an example of

    a   dormancy   b   nocturnal behaviour
    c   hibernation   d   migration

(j) A physical change that occurs only in girls during adolescence is

    a   becoming moody   b   hair growing in genital regions
    c   menstruation starting   d   voice becoming deeper   (10)

2   Look at the dental plans of teeth for a human and a dog below.

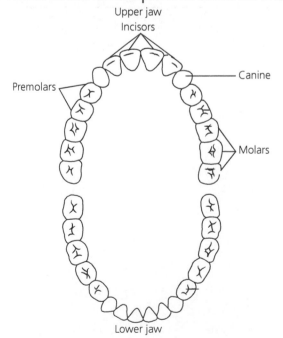
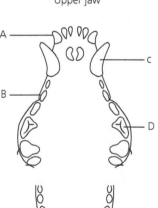

(a) Using the dental plan for a human as a guideline complete the table below for the dog.

Tooth	Name of tooth
A	
B	
C	
D	

(4)

(b) Using evidence from the dental plans, describe and explain two ways in which the diet of a human is different from that of the dog.

_____

_____

_____

_____ (4)

Both humans and dogs can have bad breath caused by bacterial action in the mouth.

(c) Describe three things you can do to make sure that you have good dental hygiene.

_____

_____ (3)

3   The picture below shows a 'one-man band'.

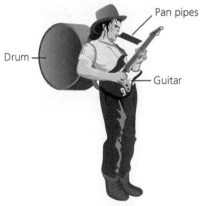

Pan pipes

Drum

Guitar

Sound is made when objects vibrate.

(a) State what is vibrating for each of the instruments in the picture.

drum: _____

pan pipes: _____

guitar: _____ (3)

Turn over to the next page.

(b) Describe how the man could make the sound louder when hitting the drum.

_____ (1)

(c) Describe two ways in which the man could play a higher note on the guitar.

_____

_____ (2)

(d) What would happen to the volume of the sound if you walked away from the 'one-man band'?

_____ (1)

(e) Describe as fully as you can how the sound travels from the instruments to your brain.

_____

_____

_____

_____ (3)

An old man complains that he cannot hear the high notes when the man is playing the pan pipes.

(f) Suggest why a young person can usually hear higher notes than an old person.

_____

_____ (2)

4   The diagram below shows the water cycle.

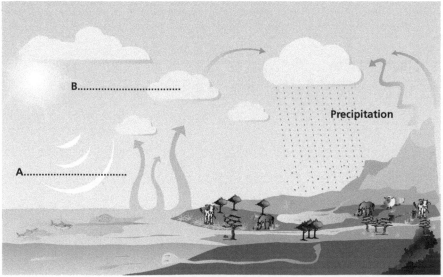

Precipitation

B.............................

A.............................

(a) Complete the labels A and B on the diagram to indicate the processes occurring in these parts of the water cycle. (2)

(b) (i) Name the source of energy shown in the diagram necessary for process A.

_____ (1)

(ii) Suggest another source of energy not shown in the diagram for process A.

_____ (1)

The water cycle works as water changes state from liquid to gas and back again.

(c) Label the diagram below to give the names of the processes where liquid changes to solid and back again.

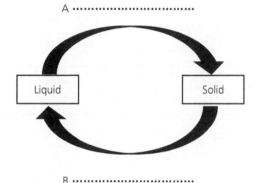

(2)

(d) Complete the boxes below to show the arrangement of particles in a solid, a liquid and a gas.

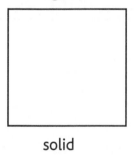  

    solid   liquid   gas (3)

5 The table below shows the changes in mass of a fetus as it grows in the uterus, in average values.

At 8 weeks the fetus has a mass of only 1 g.

Age, in weeks	Mass, in g
10	4
15	70
20	300
25	600
30	1300
35	2400
40	3500

The table shows the mass of the fetus every 5 weeks.

Turn over to the next page.

(a) You are going to show these results as a graph.
  (i) Choose a scale for the horizontal axis and add this to your graph. (1)
  (ii) Plot the data from the table on the graph. (3)
  (iii) Draw a smooth curve to fit most of the data points. (1)

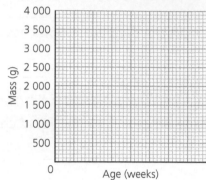

(b) Suggest why there is no data at 5 weeks.

_____ (1)

(c) Describe how the mass of the fetus changes:

from 0 to 20 weeks: _____

_____ (1)

from 20 to 40 weeks: _____

_____ (1)

This is a labelled scan of a baby at 20 weeks.

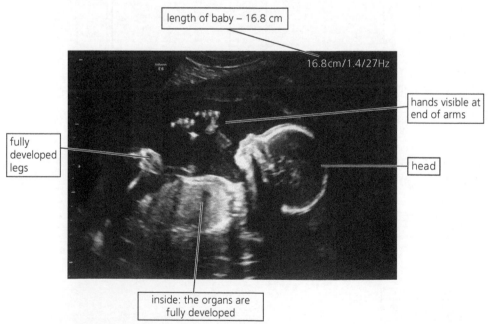

(d) Use the information in the scan to suggest how the baby develops in the first 20 weeks.

_____

_____ (2)

Amy and Hassan were both born at 40 weeks and have been told that their birth masses were 4000 g and 3000 g.

(e) Suggest why their masses at birth are different from that in the table.

_____ (1)

6 Choose options from the box to complete the sentences about the properties of materials.

absorbent	a closed switch	an open switch	melt
brightly	buzz	dimly	electrical conductor
electrical insulator	graphite	parallel	series

(a) Copper is used for the wires in an electric circuit because it a good

_____. (1)

(b) Plastic is used for the outside of a plug because it is a good

_____. (1)

(c) When there is too much current in a circuit the fuse will

_____. (1)

(d) A non-metal that conducts electricity is _____. (1)

(e)

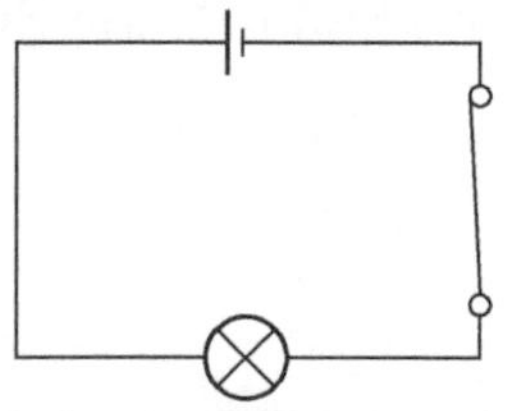

The components in the above circuit are connected in _____. (1)

(f) If you add an extra cell to the circuit in part (e) the bulb will shine more

_____. (1)

(g) The components in the circuit in part (e) are a cell, a bulb and

_____. (1)

Turn over to the next page.

7   Read the following and then answer the questions below.

> Chemists are often involved in research to develop new materials.
>
> In 1970, a chemist called Spencer Silver developed a new adhesive but it was very weak so it was ignored.
>
> In 1974, another scientist called Arthur Fry found that his bookmarks in his hymn books kept falling out.
>
> He remembered Spencer Silver's adhesive and tried it on his bookmarks.
>
> He had discovered a marker that would stay in place, yet lift off without damaging the pages.
>
> Sticky notes are now one of the most popular office products on the market.

(a) Suggest why it is important to always record your results even if your experiment has not worked.

_____

_____ (2)

(b) What were the properties of the glue that made it useful as a bookmark?

_____

_____ (2)

Spencer Silver originally wanted to use his glue as a surface on bulletin boards to which temporary notices could be attached.

(c) Suggest why this would not have been a good idea.

_____ (1)

You are asked to investigate the strength of different types of glue.

You are given a strip of strong plastic sheet with a hole at the bottom which can be used to attach the hook of the weight holder.

You are allowed to use the flat side edge of a table in your investigation.

Strip of strong plastic with hole at the bottom

Weight carrier and several 10 g weights

(d) Using the apparatus on the previous page, describe how you would test the different types of glue.

_____

_____

_____

_____ (4)

(e) How would you make your investigation a fair test?

_____

_____

_____

_____ (3)

8   Complete the table below to show the colour of litmus paper in different solutions.

Substance	Acid, alkaline or neutral	Blue litmus paper	Pink litmus paper
lemon juice	acid		
salt water	neutral		
sink cleaner	alkaline		
white vinegar	acid		

(8)

Record your results and move on to the next paper.

Score [ ] / 80   Time [ ] : [ ]

Paper 8: 11+ Practice Paper

1 Underline the option that best completes each of the following sentences.

(a) The life process that is more difficult to observe in plants is

   a growth   b movement   c nutrition   d reproduction

(b) The part of the plant that transports water from the roots is the

   a flower   b leaves   c roots   d stem

(c) An animal that is an invertebrate is

   a a bat   b an earthworm   c an eel   d a snake

(d) When the eardrum is perforated (burst) after hearing a loud explosion, damage to hearing is usually

   a negligible   b painless   c permanent   d temporary

(e) The best method to separate dry rice from a mixture of rice and salt is

   a decanting   b evaporation   c filtration   d sieving

(f) On a sunny day in summer your shadow will be longest at

   a 6:00 a.m.   b 12:00 a.m.   c 3:00 p.m.   d 11:00 p.m.

(g) When water is cooled from 20 °C to −5 °C it changes from

   a gas to liquid   b gas to solid   c liquid to solid   d solid to liquid

(h) In a mixture of chalk and water the chalk is

   a insoluble   b soluble   c the solute   d the solvent

(i) You can identify rocks from a volcano by the presence of

   a crystals   b fossils   c grains   d layers

(j) When light hits a mirror it is

   a absorbed   b bent   c captured   d reflected (10)

2 (a) On the diagram of the digestive system below label the mouth, tongue, esophagus, stomach, small and large intestine.

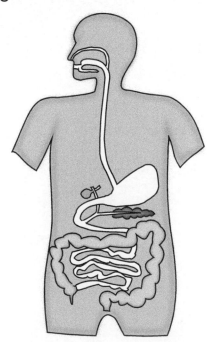

(6)

(b) Using straight lines, connect the boxes below to match the part of the digestive system with its function.

Tongue		Strong muscular bag that adds acid and churns up food
Stomach		Long thin tube that digests food
Small intestine		Strong muscle that moves food around and pushes food into esophagus
Large intestine		Wide tube that absorbs water and makes faeces

(4)

3   Helen and Lizzie were on a camping holiday and wanted to use some water from a stream for drinking.

The water had some mud in it.

They had brought a filter funnel, beaker and filter paper with them.

(a) In the space below draw a labelled diagram of how they would set up their apparatus.

(4)

Turn over to the next page.

(b) (i) The term used for the liquid that has passed through a filter is the

_____.

(1)

(ii) The term used for the solid that stays on the filter paper is the

_____.

(1)

They have been told to boil the water to make it safe to drink.

They had a thermometer and found that the water boiled at 103 °C.

They expected the water to boil at 100 °C.

(c) Suggest why the water boiled at a higher temperature than they expected.

_____ (1)

They drank a very small amount of the water and it tasted horrible.

(d) Suggest why the water tasted horrible.

_____ (2)

Fortunately for Helen and Lizzie the water was not dangerous to drink.

Water sometimes contains toxic materials such as pesticides and fertilisers that have entered the water from surrounding streams as well as disease-causing micro-organisms.

(e) Suggest one effect of these toxic materials in the stream.

_____ (1)

4 Class 6G had been making paper aeroplanes. They decided to investigate if the paper they used made any difference to how far the planes travelled.

(a) Describe how they could make their investigation a fair test.

_____

_____

_____

_____ (5)

They recorded their results.

Throw	Distance thrown, in m			
	Exercise book paper	Photocopy paper	Sugar paper	Card
1	2.0	2.0	3.0	5.0
2	2.5	3.0	4.0	3.5
3	1.0	2.5	3.0	3.5
4	1.5	2.5	3.0	4.0
5	3.0	2.5	2.0	4.0
Average	2.0	2.5	3.0	

(b) (i) Why did Class 6G repeat their experiment five times for each material and average their results?

_____ (1)

(ii) Calculate the missing average and write it in the table. (1)

(c) On the grid below:
(i) add the missing label on the vertical axis (1)
(ii) complete the bar chart by drawing in the three remaining bars. (2)

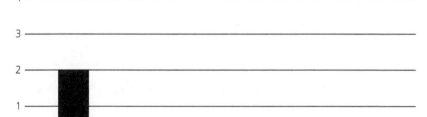

(d) (i) What force is acting to resist the movement of the plane?

_____ (1)

(ii) Suggest an explanation for their results

_____ (1)

Turn over to the next page.

5 Zara was separating salt from salt solution.
Her method is shown below.

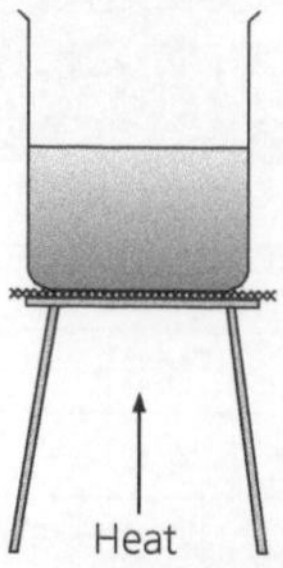

1. Heat the salt solution in a beaker using a Bunsen burner, tripod and gauze.

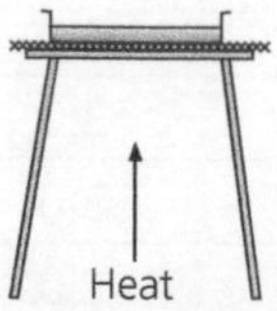

2. When the solution is reduced to about 1 cm depth transfer to an evaporating basin and continue heating.

(a) Complete the sentence below:

When the salt solution is heated the liquid _____ to become a

_____. (2)

Zara was told that salt solutions should not be dried completely when heated.

(b) Suggest why she was given this advice.

_____ (1)

(c) Suggest why Zara changed from a beaker to an evaporating basin to continue evaporating the water from the solution.

_____

_____ (2)

(d) Suggest and explain two steps that Zara should take so that she is safe during this experiment.

1. _____ (2)

2. _____ (2)

6 In the Dr Doolittle stories, there is an animal called the pushmi-pullyu.
In force diagrams, arrows of the same length represent forces of the same strength.
Long arrows show larger forces than the forces represented by shorter arrows.
Movement will be in the direction of the largest force.
Both ends of the pushmi-pullyu are equally strong.

(a) (i) Add equal length arrows to the picture to show each end of the pushmi-pullyu, trying to move in the direction in which it is facing. (2)

(ii) Will the pushmi-pullyu be moving? _____ (1)
Explain your answer.

_____

_____ (2)

(iii) Suggest the best way for the pushmi-pullyu to move forward in one direction.

_____

_____ (2)

All forces are pushes, pulls or a combination of both.

(b) Tick a box in the table below to show which forces are acting for each of the examples.

Example	Push	Pull	Both
pedalling a bike			
opening a box			
spinning a coin			
shutting the curtains			
closing a cupboard door			
playing on a swing			

(3)

Turn over to the next page.

7   Read the following passage and then answer the questions below.

Carl Linnaeus was born in 1707, in Sweden.

Before Linnaeus, living things had names that could be made up of as many as ten Latin words strung together!

As a result of his work, living things are divided into separate kingdoms: animals, plants, bacteria, fungi and single-celled organisms. The living things in each of these kingdoms have characteristic features.

Each kingdom is then divided into smaller groups. For example the animal kingdom is divided into vertebrates and invertebrates. These groups are then divided again and again until you have the name of a single living thing.

Living things now have a much simpler two-part Latin name. Humans are *Homo sapiens*.

(a) Suggest one reason why studying plants and animals would have been difficult for scientists before Linnaeus.

_____ (1)

(b) Suggest why fungi are in a separate kingdom.

_____ (1)

(c) Using information in the passage, complete the following classification chart.

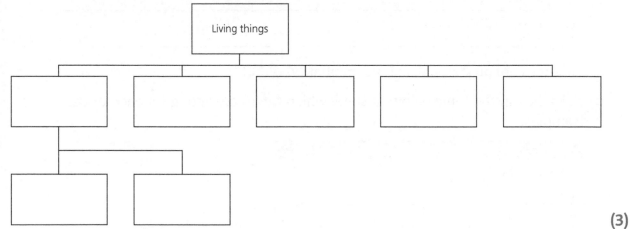

(3)

Vertebrates are divided into five groups.
In vertebrates, the type of skin is characteristic of each of these groups.

(d) Complete the table below about each group of vertebrates.

Vertebrate group	Type of skin
	hair/fur
birds	
	wet scales
	moist skin
reptiles	

(5)

8 Virtually all materials are made through chemical change.
   In each of the examples below underline the word that shows the type of chemical change that is occurring.

   (a) A candle burning as a luminous source.

   A bicycle rusting when left out in the rain.

   Mixing sand, cement and water which set to form concrete.

   Mixing eggs, sugar, butter and flour and baking to make a cake.

   Strawberries ripening and changing colour from green to red.   (5)

   Pollution is often caused by humans.

   (b) (i) Describe and explain one chemical change caused by humans that pollutes the environment.

   _____

   _____ (2)

   (ii) Suggest a way in which humans pollute the environment without chemical change.

   _____ (1)

   Sometimes pollution of the environment can occur without any involvement of humans at all.
   (iii) Suggest one way that the environment can be polluted without human involvement.

   _____ (1)

Record your results and move on to the next paper.

Score ⬚ / 80 Time ⬚ : ⬚

Paper 9: 11+ Practice Paper

1   This question is about the human body.
   (a) Complete the table below naming the organs A, B, C, D and E in the diagram.

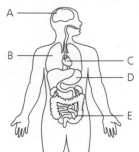

	Organ
A	
B	
C	
D	
E	

(5)

   Humans and other animals have a skeleton.
   (b) Describe the three main functions of the skeleton.

   _____

   _____

   _____ (3)

   (c) Complete the following paragraph about parts of the skeleton.

   Organ A is inside the _____ and organ B is inside the

   _____. (2)

2   Sarah and Suki have been using litmus paper to test a range of colourless liquids to see
   if they are acid or alkaline.

	Colour in		
	Acid solution	Pure water	Alkaline solution
pink litmus paper	pink	pink	blue
blue litmus paper	pink	blue	blue

   (a) Describe how you would use litmus paper to identify a neutral liquid.

   _____

   _____ (2)

   Their teacher told them that litmus was a natural substance extracted from a
   plant-like organism called lichen.
   They wondered if other plant extracts would change colour in acid and alkaline
   liquids and decided to investigate.

They obtained some plant extracts and tested them in pure water and in colourless solutions of an acid and an alkali. Their results are shown below.

	Colour in		
	Acid solution	Pure water	Alkaline solution
red cabbage water	dark pink	purple	blue/green
red onion water	red	violet	green
blackcurrant juice	bright red	purple	green

(b) The name for any material that changes colour and can be used to identify whether substances are acidic or alkaline is

_____. (1)

(c) Suggest why Sarah and Suki decided to use brightly coloured plant extracts.

_____ (1)

All of the plant extracts showed similar colours in each of the three solutions.

(d) What does this suggest about the coloured pigments in the three extracts?

_____ (1)

Their teacher told them that some red wine had been spilt on a white carpet at a party.
She had tried to clean it up with some washing soda dissolved in warm water.
The stain turned from red to violet to green.
Washing soda is alkaline.

(e) Using your knowledge about acids and alkalis suggest what was happening to the stain as the teacher tried to clean it up.

_____

_____

_____ (3)

3 Look at the diagrams below of five different pairs of circuits.
In each case complete the sentence below.

(a)

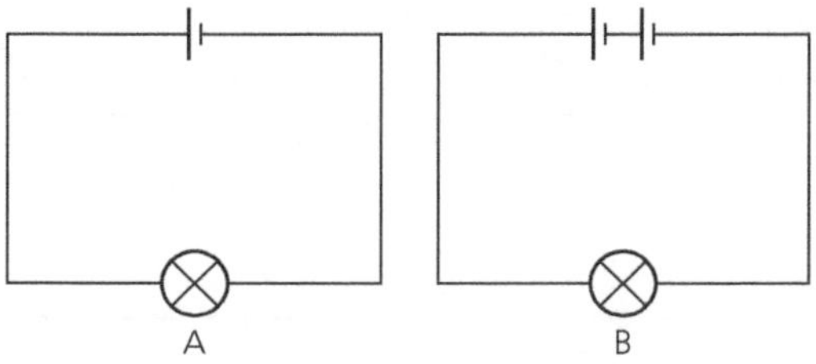

The bulb in circuit B is _____
compared with the bulb in circuit A

because _____. (2)

Turn over to the next page.

(b)

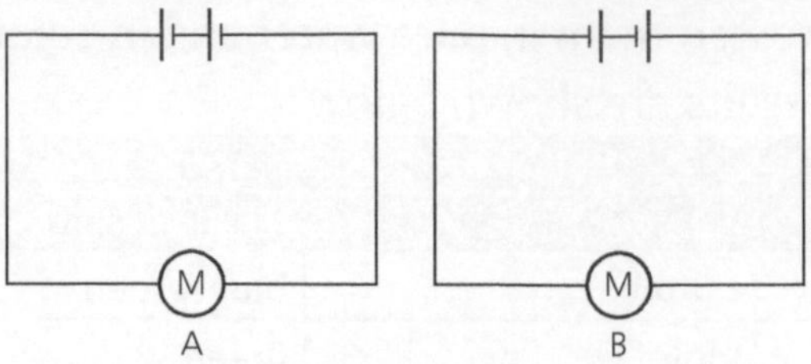

The motor in circuit B turns _____
compared with the motor in circuit A

because _____. (2)

(c)

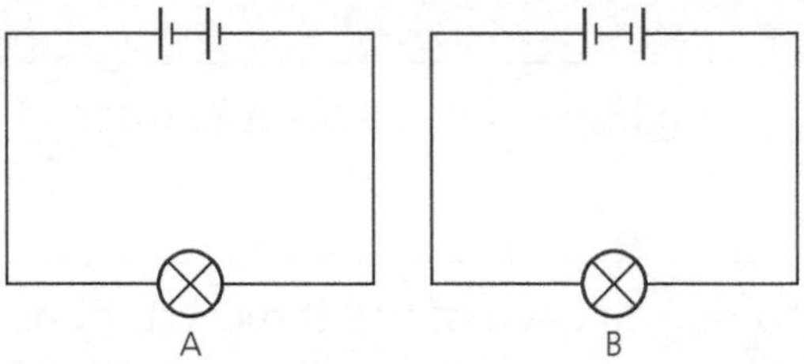

The bulb in circuit B is _____
compared with the bulb in circuit A

because _____. (2)

(d)

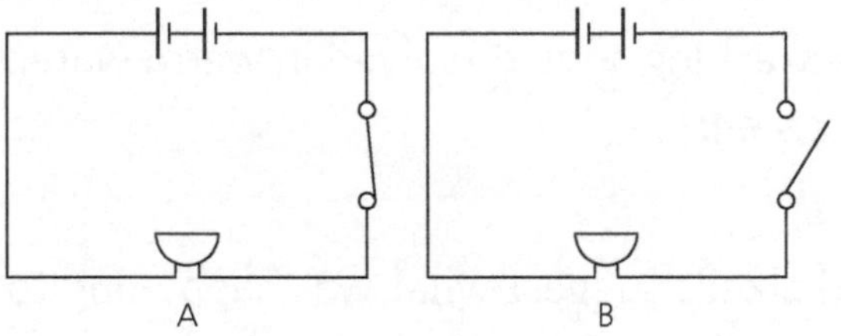

The buzzer in circuit B is _____
compared with the bulb in circuit A

because _____. (2)

(e)

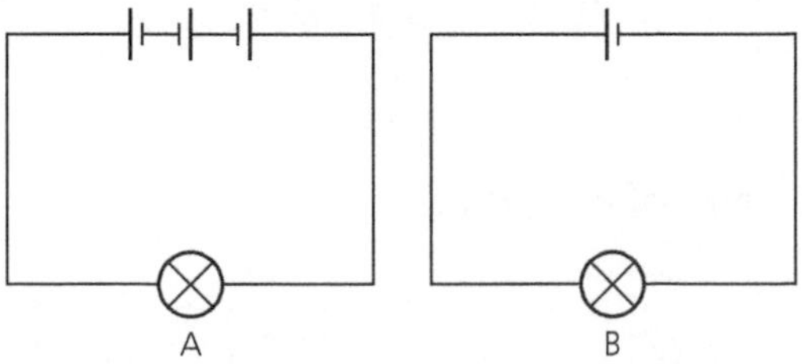

The bulb in circuit B is _____
compared with the bulb in circuit A

because _____

_____. (2)

4   The students in Year 6 have been studying insulation.

They were investigating different thicknesses of insulating materials.

They used a glass beaker and wrapped the insulating materials around it.

They put the same volume of hot water into the beaker each time and measured the temperature every 20 minutes.

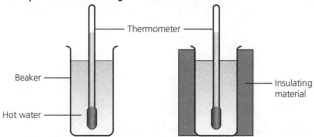

They started the experiment when the water temperature was 60 °C.

An investigation needs a suitable question as its starting point.

(a) (i) Write below a suitable question for this investigation.

_____ (1)

   (ii) Which variables did they control to make sure that it was a fair test?

_____

_____ (3)

   (iii) Suggest why it was a good idea to use a tall, narrow beaker rather than a short, fat beaker.

_____ (1)

They started their investigation by measuring the fall in temperature of the water when there was no insulation.

They plotted their results as a line graph as shown here.

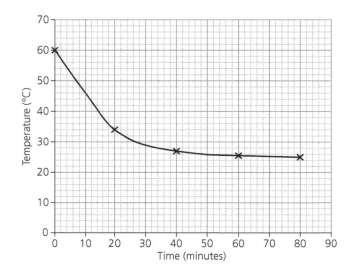

(b) What would be the temperature of the water after 100 minutes?

_____

Explain your answer.

_____

_____

_____ (2)

Turn over to the next page.

They then wrapped the beaker with 2 cm thickness of cotton wool and obtained the following results.

Time, in minutes	Temperature, in °C
0	60
20	48
40	38
60	32
80	28

(c) (i) Plot these results on the graph on the previous page. (2)

(ii) Join your points with a smooth curve. (1)

They proceeded to investigate different thicknesses of cotton wool.

(d) Describe the curve you would predict if you added 4 cm thickness of cotton wool as an insulating layer.

_____

_____ (2)

5 Most animals reproduce sexually and all plants can reproduce both sexually and asexually.

(a) Complete the following passage about reproduction in animals and plants using words from the box.

You may use words once, more than once or not at all.

carpel	dispersal	fertilisation	fusion	
germination	identical	meeting	similar	stamen

Sexual reproduction in plants and animals involves the _____ of

male and female sex cells in a process called _____. In plants,

the male sex cells are contained in the _____ and transferred

to the female sex cells in the _____ in a process called

pollination. Sexual reproduction produces offspring with characteristics that vary

from the characteristics of the parent animal or plant.

Some microscopic animals reproduce asexually by splitting into two when they reach

a certain size. Plants can reproduce asexually by a variety of methods. The offspring

are _____ to the parent plant. Plants often use both methods

of reproduction. (5)

(b) Read the following descriptions and show by ticking the correct box whether the organism uses sexual, asexual or both methods of reproduction.

Description	Sexual	Asexual	Both
A strawberry plant has small white flowers and has runners that grow into new plants.			
An amoeba consists of one cell that divides into two when it gets large.			
An oak tree uses the wind for pollination and squirrels for seed dispersal.			
A pair of rabbits could produce up to 96 baby rabbits in a year.			
A hydra is a tiny freshwater animal about 1 cm long. It reproduces by budding where the offspring grow out of the body of the parent.			

(5)

6 Look at the diagram of a human sundial below.

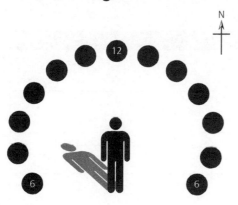

(a) At what time of day will the shadow be the shortest? _____
Explain your answer.

_____ (2)

(b) Is the time shown on the sundial in the morning or afternoon? _____
Explain your answer.

_____

_____ (3)

(c) Why are there no markers directly behind where the person is standing?

_____ (1)

(d) Complete the following sentences about the movement of the Earth, Sun and Moon.

(i) We have day and night because _____. (1)

(ii) We have years because _____. (1)

(iii) We have a full moon approximately once every 28 days because

_____. (1)

Turn over to the next page.

7 The diagram below is a very simple key to identify rocks.

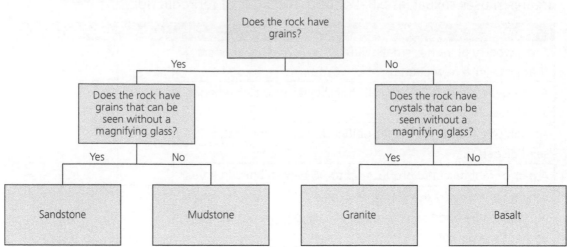

(a) Use the key to identify the following rocks from the descriptions you have written in your field notebook.

Field notebook entry	Rock
black rock that needed a hand lens to show that it was made of tiny crystals	
yellow/brown rock that was made up of clearly visible grains	
grey rock that needed a hand lens to show that it was made of tiny grains	
grey rock that was made up of clearly visible large crystals	

(4)

(b) Which feature written in the field notebook was not used by the key to identify the rocks?

_____ (1)

You have two different rocks made of clearly visible grains. You have recorded the following about them.

Rock 1: yellow/brown in colour, clearly visible large grains, crumbly

Rock 2: yellow/brown in colour, clearly visible small grains, contains fossils

It has been suggested that you add the following question to replace 'sandstone' in its box in the key.

Are the grains large?

(c) (i) Suggest why this might not be a good question to have in a key.

_____

_____ (2)

(ii) Write an alternative question that can be used to identify these two rocks as part of this key.

_____ (1)

The key names two sedimentary rocks, sandstone and mudstone.

(d) Name a different sedimentary rock that frequently contains fossils.

_____ (1)

8 Read the passage below and then answer the questions.

> The Wright brothers are credited with inventing the aeroplane.
>
> On 14th December, 1903, they made the first successful manned flight in an aeroplane, powered by an engine, that was heavier than air.
>
> The first flight lasted only 12 seconds and the aeroplane flew for just over 36 metres.
>
> By the following year they had increased the length of time in the air to over five minutes.
>
> They had to invent lightweight engines and propellers. They observed how birds flew and this helped them to design the wings for their aeroplanes.

(a) Label the forces on the diagram below.

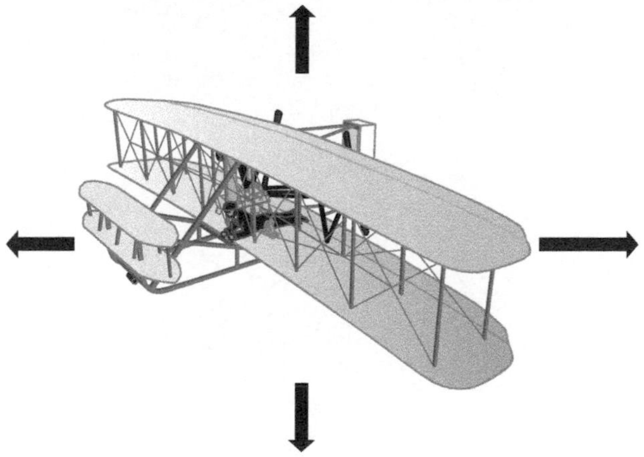

(4)

Large forces are shown using longer arrows than those used for smaller forces. Movement is in the direction of the largest force.

(b) (i) Add a labelled arrow to the diagram to show the direction of movement of the aeroplane. (2)

(ii) Explain your answer.

_____ (1)

It is the design of the wings that creates the upward force on the aeroplane.

(c) How did the Wright brothers get the idea for the design for the wings?

_____ (1)

Turn over to the next page.

9 Look at the food chain below.

| cabbage ⟶ caterpillar ⟶ sparrow ⟶ hawk |

(a) What do the arrows represent in the food chain?

_____ (1)

(b) (i) What is the original source of energy for the food chain?

_____ (1)

(ii) Explain how this original source of energy enters the food chain.

_____ (2)

Record your results and move on to the next paper.

Score ☐ / 80 Time ☐ : ☐

Paper 10: 11+ Practice Paper

1 Life processes include nutrition, movement, growth and reproduction.

(a) Tick the appropriate boxes to show the processes being described in each of the examples in the table.

In some of the examples there may be more than one life process is being described.

Example	Nutrition	Movement	Growth	Reproduction
a lion chasing a gazelle				
green leaves on a plant absorbing energy from the Sun				
digested foods in an animal being used to build new cells				
a flower producing seeds, which then germinate to produce new plants				
a rabbit running away from a predator hawk				

(6)

Plants cannot move from place to place in same way as animals but leaves will often turn to face the Sun.

(b) Suggest why leaves turn to face the Sun.

_____

_____ (2)

Look at the picture below.

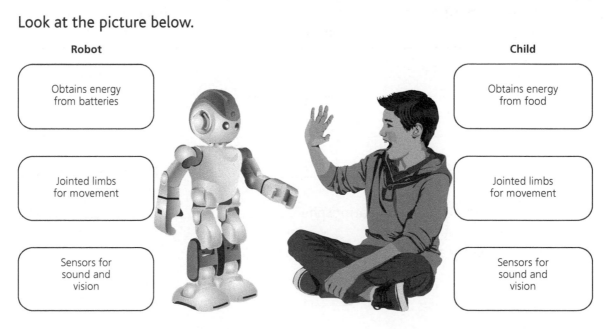

Robot

Obtains energy from batteries

Jointed limbs for movement

Sensors for sound and vision

Child

Obtains energy from food

Jointed limbs for movement

Sensors for sound and vision

Turn over to the next page.

The robot and the child have some life processes in common.

(c) Explain, using your knowledge of life processes, why the child is a living thing and the robot is not.

_____

_____ (2)

2   The diagram below shows the Earth, the Sun and the Moon.

Moon

Sun

A

Earth

(a) (i) Draw a ray of light to show how light travels from the Sun to the Moon.   (2)
     (ii) Draw a ray of light to show how a person standing at A on the Earth sees
          the Moon.   (2)

(b) Complete the following paragraph about the Sun, Earth and Moon using words from the box below.
    Each word can be used once, more than once or not at all.

direct	light	luminous	Moon
non-luminous	rays	reflected	
Sun	transparent	translucent	

The Sun is a _____ object because it gives out

_____. The Moon is a _____ object and

can only be seen by _____ light. On Earth we can see the Sun

and the Moon because the atmosphere is _____. It is dangerous

to look directly at the _____.   (6)

3   Sand, clay and loam are three main types of soil.

   (a) Using straight lines, match each soil to the box containing the text that best describes it.

Sand	Dark in colour, sticky when wet, can be moulded to form a sausage shape
Clay	Dark in colour, slightly gritty texture, feels moist
Loam	Light in colour, gritty, feels dry, grains fall through fingers

(3)

   The descriptions suggest that some soils hold water better than others.
   You have been asked to investigate the drainage properties of each soil.
   You have been provided with the following:

   > three filter funnels and paper
   > beaker with 100 cm³ line marked
   > clock
   > weighing scales
   > three 100 cm³ measuring cylinders
   > samples of each type of soil – sand, clay and loam

   (b) Complete the four missing stages of the method below for your investigation, making sure that it is a fair test.

   1. Set up three filter funnels with filter paper and place measuring cylinders beneath.

   2. Weigh equal masses of each sample of soil and put in filter funnels.

   _____

   _____

   _____

   _____ (3)

Turn over to the next page.

(c) (i) Predict which soil will drain the fastest. _____ (1)

(ii) Explain why you have made this prediction.

_____

_____ (2)

The loam soil is dark in colour because it contains organic matter.

(d) (i) What is the name given to this organic matter? _____ (1)

(ii) Suggest and explain one reason why farmers and gardeners generally prefer loam soils for growing their plants.

_____

_____ (2)

4 Pesticides are chemicals that kill insects, weeds and the small animals that eat crops. Unfortunately, they also kill organisms that are harmless.

Organic farmers do not use pesticides.

(a) Explain why farmers often spray pesticides on their crops.

_____

_____ (2)

Rachel Carson was born in 1907 in the USA.

She discovered that a pesticide called DDT harmed the environment and made people sick.

She wrote a book called *Silent Spring* so-named because of the effect that DDT was having on birds. If all of the birds died, there would be no birdsong and springtime would be silent.

(b) Suggest two possible advantages of organic farming.

1. _____. (1)

2. _____. (1)

Birds did not die directly from the effects of the DDT but it made the shells of their eggs very thin.

(c) Suggest why thin eggshells could lead to a decrease in bird populations.

_____

_____ (2)

The bald eagle, the national bird of the USA, nearly became extinct partly as a result of the use of DDT.

DDT was washed into rivers and streams where the microscopic plants and animals that lived there absorbed it.

These microscopic plants and animals were eaten by small fish, which absorbed the DDT.

The small fish were eaten by large fish.

The large fish were eaten by birds such as the bald eagle.

The DDT levels increased at every stage of the food chain.

(d) Use this information to complete the food chain below.

(2)

5 Sarah has got a baby sister and she complains that the baby's crying is making her ears hurt. She used a sound meter and measured the sound of her sister crying at different distances from her.

She obtained the following results.

Distance, in m	Sound level
0	140
1	104
2	94
3	90
4	89
5	88
6	87

Turn over to the next page.

(a) On the grid:

    (i) label the vertical axis   (1)

    (ii) add a scale to the horizontal axis   (1)

    (iii) plot the points in the table   (3)

    (iv) join the points with a smooth curve.   (1)

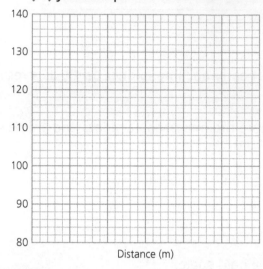

Distance (m)

(b) Describe the pattern of results shown by the graph.

_____

_____ (2)

Sarah looked at the chart of sound levels in her classroom.
It looked like this:

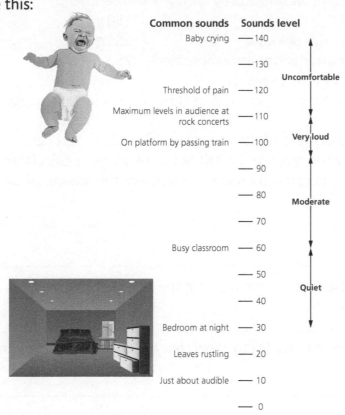

Common sounds	Sounds level	
Baby crying	— 140	
	— 130	Uncomfortable
Threshold of pain	— 120	
Maximum levels in audience at rock concerts	— 110	
On platform by passing train	— 100	Very loud
	— 90	
	— 80	Moderate
	— 70	
Busy classroom	— 60	
	— 50	Quiet
	— 40	
Bedroom at night	— 30	
Leaves rustling	— 20	
Just about audible	— 10	
	— 0	

(c) Using information from your graph and the chart, was Sarah justified in saying that the sound made her ears hurt?

_____

_____ (2)

6　When you are deciding how different materials can be separated from a mixture, knowledge of the properties of materials is important.

(a) For each of the following, suggest a method to separate the material from its mixture and explain why your method works.

(i)　Iron filings from a mixture of iron filings and sand

Method: _____ (1)

The method works because _____

_____ (2)

(ii) Chalk powder from a suspension of chalk and water

Method: _____ (1)

The method works because _____

_____ (2)

Turn over to the next page.

(b) In an investigation, a sample of seawater was evaporated slowly in a beaker.
The following data was measured during the experiment.

mass of beaker 250 g
mass of beaker with seawater 350 g
mass of beaker after water has evaporated 253 g
mass of the salt in the seawater 3 g

There is a space for your workings in each of the following questions.

(i) Calculate the mass of the seawater. _____ (1)

(ii) Explain how the mass of the salt in the seawater is calculated.
You can just show the calculation used to arrive at the answer of 3 g if this is easier.

_____

_____ (1)

(iii) Calculate the mass of the water in the seawater mixture.

_____ (1)

(c) Use your data from part (b) to show that mass is conserved during physical changes.

_____

_____ (1)

(d) Evaporation was used to separate the salt from the seawater.

The method works because _____

_____ (2)

7 (a) In the space below draw a circuit diagram with two cells, a closed switch and a buzzer connected in series.

(5)

Archie connected his circuit incorrectly and discovered that the cells became very hot and the buzzer did not work.

(b) Suggest what Archie may have done incorrectly and explain why this can be dangerous.

_____

_____ (2)

8   For good health it is important to eat a balanced diet.

(a) What does the phrase 'balanced diet' mean?

_____

_____ (2)

The following nutritional information is from a ready meal of spaghetti bolognese.

	Per serving of 400 g, in g
protein	32
carbohydrate	60
fat	20
fibre	4
salt	1
total	

(b) (i) Add up the data in the table above and fill in the total.   (1)
    (ii) Suggest which material makes up the rest of a 400 g serving.

    _____   (1)

The chart below shows recommended amounts of food that should be eaten by an 11-year-old each day.

	Food per day, in g
protein	45
carbohydrate	320
fat	90
fibre	24
salt	6

(c) Look at the two charts and complete the comments column of the table below to indicate how a ready meal of spaghetti bolognese can be a part of your daily diet. Protein has been done for you.

	Comments
protein	The ready meal provides 32 out of the 45 g of protein that I need so I do not need much more protein today.
carbohydrate	
fat	
fibre	
salt	

(4)

Turn over to the next page.

You are planning to have the ready meal of spaghetti bolognese for your lunch.

(d) Tick the option, A or B, for each of the other meals you could eat so that, overall, you will have a balanced diet today.

Breakfast:
A Two sausages, scrambled egg, fried bread and beans ☐
B Cereal with milk, two slices of wholemeal toast and butter,
   orange juice ☐

Dinner/supper:
A Baked potato with some grated cheese and beans with salad ☐

B Steak, chips, peas and onions ☐

Snacks:
A Cake, a large packet of salt and vinegar crisps and fizzy drinks ☐

B Jam sandwich with a glass of blackcurrant squash ☐   (3)

Record your results and move on to the next paper.

Score ☐ / 80   Time ☐ : ☐

Paper 11: 11+ Practice Paper

Test time: 60:00

1 Underline the option that best completes each of the following sentences.

(a) The part of a flowering plant which carries food to all parts of the plant is the
   a flower   b leaves   c root   d stem

(b) The root of a plant does not
   a anchor the plant   b make food   c take in minerals   d take in water

(c) A type of force that can act at a distance is
   a friction   b magnetism   c upthrust   d weight

(d) In the water cycle the formation of snow is an example of
   a condensation   b evaporation   c freezing   d precipitation

(e) An example of a reversible change is
   a burning   b concrete setting c dissolving   d rusting

(f) The image below is a

   a gear   b lever   c pulley   d wheel

(g) We can see the Moon and the planets because they
   a absorb light   b are luminous   c are non-luminous d reflect light

(h) A bee is classified as an insect because it has
   a an internal skeleton b fur   c three body parts d wings

(i) The unit of force is represented by the symbol
   a F   b G   c N   d W

(j) The picture below is a trilobite fossil.

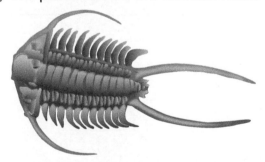

The trilobite no longer exists and is best described as
a a living thing trapped in sedimentary rock
b a non-living thing trapped in sedimentary rock
c a once-living thing preserved in sedimentary rock
d a shell preserved in sedimentary rock

(10)

Turn over to the next page.

2   Year 4 students have been looking at how things can be grouped in a variety of ways. They were given the following objects from the nature corner in their classroom.

Pet guinea pig

A cactus

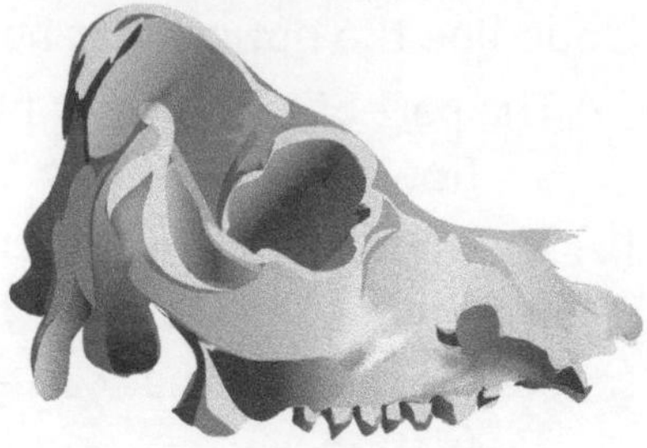

A sheep skull

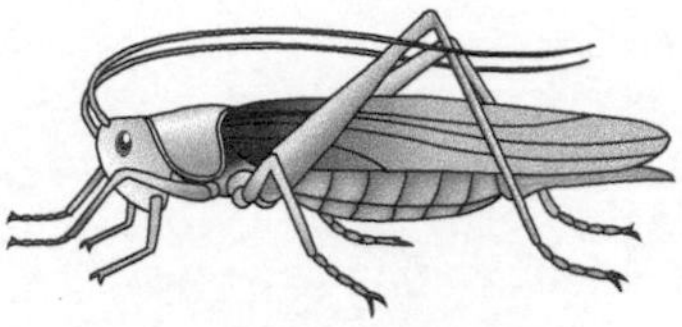

A locust

A sunflower plant

Dried flowers

They were asked to sort the objects into two groups. They discovered that they could do this using a variety of methods.

Each method needed a question with a sentence pattern like the one below:

Is it _____ or _____?

(a) Suggest two possible questions that could be used to sort the nature corner objects into two groups. Complete the tables to show appropriate headings and which objects belong in each group for your question.

Question 1: _____ (1)

group 1: _____	group 2: _____

(2)

Question 2: _____ (1)

group 1: _____	group 2: _____

(2)

The class of 15 children then decided to sort themselves into groups.
They discovered the following about themselves.

Question 1: Are you a girl or a boy?	
girl	8
boy	7

Question 2: Do you have blue or brown eyes?	
blue	5
brown	10

Question 3: Is your hair very short?	
yes	2
no	13

(b) Complete the bar chart below for these results by:
    (i) labelling the vertical axis   (1)
    (ii) plotting the missing bars.   (3)

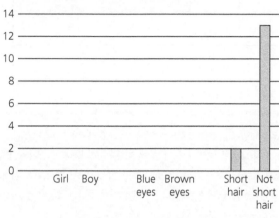

Girl   Boy   Blue   Brown   Short   Not
eyes   eyes   hair   short
hair

Grouping things is the basis of making keys.

(c) (i) Explain why the best questions in keys have answers that are either 'yes' or 'no'.

_____ (1)

    (ii) Suggest one reason why question 3 about hair length is a poor question

    to have in a key. _____ (1)

3   The drawing below shows a set of soil sieves.
    These can be used to separate particles of different sizes in soils.

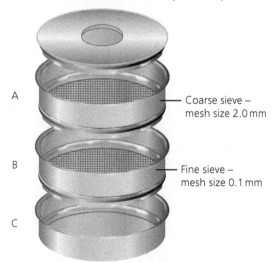

A

Coarse sieve –
mesh size 2.0 mm

B

Fine sieve –
mesh size 0.1 mm

C

Turn over to the next page.

The sieves were fixed firmly together in the order shown in the drawing on the previous page.

200 g of soil was placed in the top sieve and the lid fixed firmly on the top.

The whole set of sieves was then shaken vigorously for 5 minutes.

The amount of soil in each sieve was then weighed.

(a) Complete the first column of the table to indicate which sieve in the drawing, A, B or C, matches the grain size shown in the second column. (3)

Sieve	Particle size	Mass of soil, in g	Description of contents
	more than 2.0 mm	10	
	0.1 mm to 2.0 mm	115	
	less than 0.1 mm	75	

The contents of the sieves were described as follows:

In one sieve the soil was very stony and you could see the particles easily. It also contained some bits of wood and dead leaves. In another of the sieves the particles were very fine and I needed a hand lens to see the grains clearly. In the last sieve the particles were easily visible and felt gritty. In the last two sieves my hands became black/brown when I rubbed the soil between my fingers.

(b) Using the information above, summarise the descriptions of the contents of each sieve in the last column of the table. (3)

Particle sizes in soils can also be looked at using the sedimentation test.

In this test some soil and water are put into a jar and shaken.

The jar is then left until the soil has settled in the jar.

The drawing below shows a sedimentation test for a sample of the same soil that was put into the soil sieves.

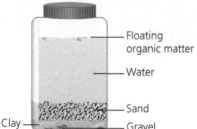

Floating organic matter

Water

Sand

Clay

Gravel

(c) What is the name given to the organic material in soil? _____ (1)

(d) Which of the materials named in the drawing of the sedimentation test would have been collected in the bottom soil sieve in the first investigation

above? _____ (1)

Soils can be simply grouped into sand, clay and loam soils.
The drawing below shows the particles in each of these types of soil.

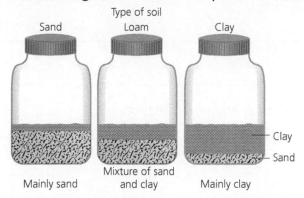

Type of soil

Sand   Loam   Clay

Clay

Sand

Mainly sand   Mixture of sand
and clay   Mainly clay

(e) (i) Suggest which type of soil has been investigated using the sieves and the

sedimentation test. _____   (1)

(ii) Explain your answer. _____

_____ (2)

4   The picture below shows a fuse connected in an electrical circuit.

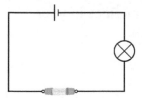

(a) Complete the following using words from the box.
Words may be used once, more than once or not at all.

breaks	clever	circular	glow
hair	makes	melt	safety
series	shocks	wire	

The metal ends of the fuse are connected to terminals as a part of a

_____ circuit. The electricity can pass through one metal end,

along a thin _____ to the other metal end of the fuse.
If too much electricity flows through the fuse the wire will get hot and

_____ and this _____ the circuit.

Fuses are _____ devices used to protect delicate electric

components and to prevent electric _____.   (6)

Turn over to the next page.

(b) Your friend is not very safety aware. Suggest two more things he should do to keep safe when working with electricity.

_____

_____ (2)

5   Water is a very unusual material because it expands on freezing.

(a) For each of the following everyday situations below suggest and explain a possible consequence if the water freezes.

(i) Water in a hosepipe on a frosty morning _____

_____ (2)

(ii) Water in the cracks of an ancient building _____

_____ (2)

Freezing water in pipes in the home can cause considerable damage.

(b) Suggest and explain a possible consequence of water freezing in pipes in your

home. _____

_____ (2)

Home owners are advised to take precautions to stop this happening and one of the things they can do is to insulate the pipes.

Look at the drawing below and then answer the questions.

Inside the house

Outside the house

Pipe insulating tape

Pipe insulation

Seal cracks and opening near pipes

(c) (i) Suggest and explain two properties that are important when choosing the material for insulating the pipes.

1. _____

_____

2. _____

_____ (4)

(ii) Suggest why gaps around the pipes where they go through an outside wall should be sealed.

_____ (1)

Home owners are advised to leave their central heating on if they go away in the winter, but they do not want to waste too much money on heating fuel. They select a temperature on the central heating thermostat before they go.

(iii) Tick the temperature below that is best for the home owners' needs. (1)

Temperature, in °C	
1	
10	
20	

6 Ann, Sofia, Zoe and Jasmin decided to grow some beans.

Ann wanted to start her bean plants early so she planted some, in early February, in pots in her greenhouse. She made sure that she watered the pots enough to keep the soil moist but not wet.

By the end of March her bean plants had still not germinated.

(a) (i) Suggest a reason for Ann's bean seeds not germinating.

_____ (1)

Her friend Sofia had had the same idea.

She planted her bean seeds in pots and added water to make the soil wet. She then wrapped the pots tightly with some plastic film to prevent water loss so she would not then have to go to the greenhouse and water the pots every few days.

By the end of March her bean plants had still not germinated.

(ii) Suggest a different reason for Sofia's bean seeds not germinating.

_____ (1)

Turn over to the next page.

Zoe decided to plant her beans in pots in the greenhouse and just leave them. By the end of March her bean plants had still not germinated.

(iii) Suggest a different reason for Zoe's bean seeds not germinating.

_____ (1)

Jasmin had meant to plant her bean seeds but had forgotten and only remembered when she was cleaning the greenhouse at the end of March.

She planted them in pots, watered them and in a week they were showing new shoots.

(b) Suggest why Jasmin's seeds germinated so quickly.

_____

_____ (3)

Jasmin liked to keep the potting bench in her greenhouse clean and tidy.

She used some wet paper towels to clean the potting bench and put the used paper towels and some beans that were left over in a large, clear jam jar to throw out later.

A week later these beans had germinated as well.

(c) What might this suggest about the needs of a bean seed for germination?

_____

_____ (2)

(d) Look at the diagrams on the previous page and describe the sequence of events in the germination of the bean seed.

_____

_____

_____

_____ (3)

7 Ptolemy was born in Egypt in the year 90.
   He believed that the Earth was a sphere and that all of the other planets moved around the Earth in circular orbits.

   (a) (i) Which of Ptolemy's ideas are still believed by scientists today?

   _____

   _____ (2)

   (ii) Which of Ptolemy's ideas is not believed by scientists today?

   _____ (1)

   Copernicus was born in Poland in 1473.

   Although his ideas were strongly opposed at the time, he believed that the Sun was at the centre of our solar system.

   (b) Suggest why his ideas might have been strongly opposed.

   _____

   _____ (1)

   We now know that Copernicus was right.

Turn over to the next page.

(c) Complete the following statements about what we know of the Earth and space today.

(i) The Sun is an example of a _____ (1)

(ii) The Milky Way is an example of a _____ (1)

(iii) The Sun, the planets and their moons all have a shape that is

_____ (1)

(iv) The force that keeps the Moon in its orbit around the Earth and

the Earth in its orbit around the Sun is _____ (1)

(d) Explain why we have day and night on Earth. _____

_____ (2)

8 Natasha was watching some dandelion seeds floating in the wind and thought that they looked like tiny parachutes.

She and her friends decided to investigate whether real parachutes stayed in the air for longer if they varied in size.

They used a plastic model of a little man to represent the seed and used cloth of different sizes for the parachute.

Using straight lines link the statement on the left to the part it plays in the investigation on the right.

Statement

Part played in the investigation

| The height that the parachute is dropped from |
| This is a prediction. |

| The size of the parachute is changed and the time for it to reach the ground is measured. All other variables are kept the same. |
| This is a fair test. |

| The time measured for the parachute to reach the ground |
| This is a control variable. |

| Natasha thought that the largest parachute would take the longest to reach the ground. |
| This is the independent variable. |

| The time was measured five times for each parachute and the results were averaged. |
| This is a method to make sure that results are reliable. |

| The size of the parachute |
| This is the dependent variable. |

(6)

Record your results and move on to the next paper.

Score ☐ / 80 Time ☐ : ☐

Paper 12: 11+ Mock Exam

Answers are to be written on the question paper. Answer all the questions.

You are allowed to use a calculator.

1   Select words from the box to complete the following sentences. Each word may be used once, more than once or not at all.

air resistance	amphibian	brain	brighter	conductor
dimmer	fossils	gravity	humus	insulator
reptile	soluble	vibrate		

(a) Sounds are made when objects _____   (1)

(b) Copper is used for household wiring because it is an electrical

_____   (1)

(c) A tadpole is a young _____   (1)

(d) Adding another bulb to a circuit makes the bulbs _____   (1)

(e) Sugar dissolves in water because it is _____   (1)

(f) The force that slows a spacecraft re-entering the atmosphere is

_____   (1)

(g) The remains of once-living things found in rocks are called

_____   (1)

2   Changes in materials may be reversible or non-reversible.

(a) Tick the correct box to show whether each of the following changes is reversible or non-reversible.

(i)   An ice lolly melting

reversible   ☐

non-reversible   ☐   (1)

(ii) A tomato ripening

reversible   ☐

non-reversible   ☐   (1)

(iii) Superglue setting

reversible   ☐

non-reversible   ☐   (1)

(iv) Water condensing on a window

reversible   ☐

non-reversible   ☐   (1)

Rusting is a non-reversible change.

(b) (i)   Which two conditions are needed for rusting to take place?

_____ (2)

(ii) Bob noticed that the metal part of his garden spade went rusty but the metal part of the greenhouse did not rust.
Suggest a reason for these observations.

_____

_____ (2)

(iii) Suggest one way Bob could prevent his garden spade from rusting in future. Explain your answer.

_____

_____ (2)

Turn over to the next page.

3 Jane Goodall is a scientist who studies chimpanzees in East Africa.

Here are some facts that she noticed about the behaviour of chimps in the wild.

- Chimps can walk along the ground and are very good at swinging through the trees.
- Chimps spend some time grooming each other's fur.
- Chimps eat fruit, leaves, insects, eggs and meat.
- Chimps usually give birth to one baby, which is fed on milk.
- Chimps use tools, for example thin sticks used to 'fish' for insects in small holes.

Use the information in the list and in the picture to answer these questions.

(a) To which animal group do chimpanzees belong? Explain your answer.

_____

_____ (3)

(b) What word would you use to describe the diet of chimpanzees?

_____ (1)

(c) Suggest two ways in which the chimp is adapted to survive in its habitat.

1. _____

2. _____ (2)

(d) The chimpanzee is an endangered animal. What does the term 'endangered' mean?

_____ (1)

(e) Suggest one way in which people could help to protect chimpanzees.

_____ (1)

4   Asha and Aled were investigating soils. They had three soil samples.

First they dried the soils. They then took 100 g of each soil and used a set of sieves to separate it according to particle size.

They recorded their results in a table.

Soil	Mass of largest particles, in g	Mass of medium sized particles, in g	Mass of smallest particles, in g
Soil A	70	20	10
Soil B	40	30	30
Soil C	20	10	

(a) Complete the table by calculating the mass of the smallest particles in soil C. Show your working.

_____ (2)

(b) Soils can be described as sand, clay or loam. Suggest which of the children's soil samples would best be described as each of these types.

   • sand _____

   • clay _____

   • loam _____   (2)

(c) Soils usually contain the remains of once-living plants and animals. What name is given to this material in the soil?

_____ (1)

(d) Asha and Aled then used the apparatus shown in the diagram to test how well water drained through each of the soils.

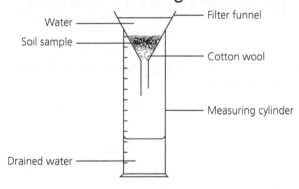

Water — Filter funnel
Soil sample —
Cotton wool
Measuring cylinder
Drained water —

Describe briefly how they might use this apparatus to find out which of the three soils allowed water to drain through fastest.

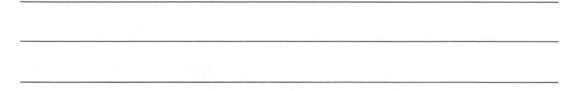

_____

_____

_____

_____ (4)

Turn over to the next page.

(e) Which of the three soils would you expect to allow the water to drain through fastest?
Explain your answer.

_____ (2)

5 Some children make four circuits. Circuit A has one bulb and one cell.
Look carefully at circuits B, C and D. Say how the brightness of the bulbs will compare with the one in Circuit A.

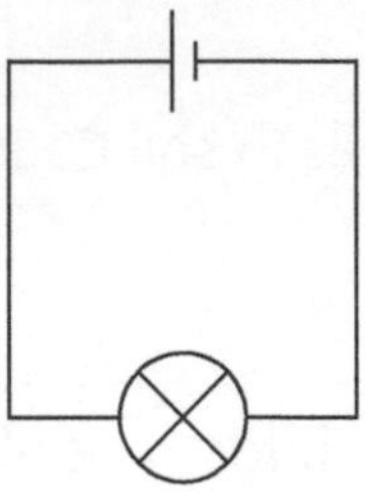

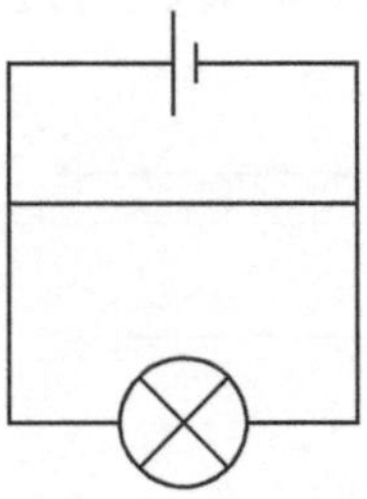

(3)

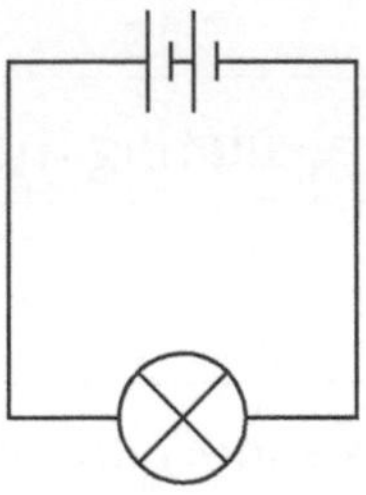

6 Robert and his mother like to grow vegetables in their garden.
They try to provide the best possible growing conditions for their plants.
(a) Give three conditions that plants need to grow well.

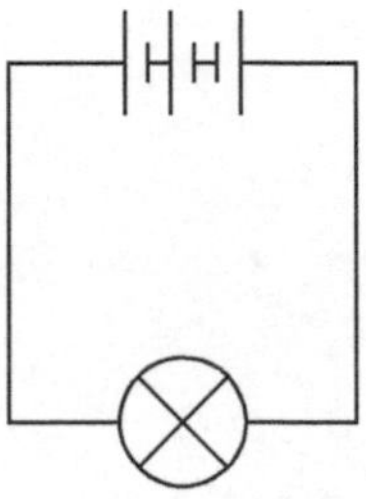

_____ (3)

(b) What name is given to the process used by plants to make their own food?

_____ (1)

Robert and his mother collect waste plant material from the garden and kitchen to make compost. They dig the compost into the soil.
(c) Give two ways in which adding the compost to the soil helps the vegetable to grow well.

1. _____

2. _____ (2)

(d) Robert knows that many vegetables grow best in alkaline soils. Suggest how Robert might carry out a test to show whether the soluble materials in their soil are alkaline.

_____

_____

_____

_____ (4)

7 (a) What temperature is shown on the following thermometers?

A _____ B _____ C _____ (3)

(b) What units should be used to measure the following items?

(i) The force needed to pull a drawer open. _____ (1)

(ii) The distance from London to Paris. _____ (1)

(iii) The volume of coffee in a mug. _____ (1)

(iv) Your mass. _____ (1)

(v) The area of the room. _____ (1)

(c) The pages of a book are too thin to measure easily. Suggest how you could estimate the thickness of one page in the book.

_____

_____ (2)

8 The New Horizons space probe flew past Pluto in July 2015. It was launched from Earth in 2006.

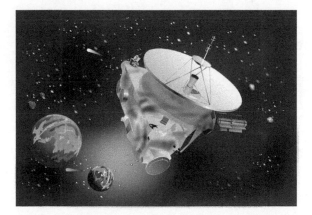

Pluto is classed as a minor planet. It is further from the Sun than all the other planets. Pluto is a little smaller than Earth's moon.

(a) Name the two planets with orbits closest to Pluto.

_____ (2)

(b) What force keeps Pluto in orbit round the Sun?

_____ (1)

Turn over to the next page.

(c) Pluto was discovered 1930 by an astronomer called Clyde Tombaugh. At the time, Pluto was considered to be a planet. All the other planets were discovered a long time earlier. Suggest two reasons why it took so long for astronomers to discover Pluto.

1. _____

2. _____ (2)

9 Some people are watching a shadow puppet play.

(a) The puppet makes a shadow on the screen. Explain how a shadow is made.

_____

_____ (2)

(b) The puppet is moved further from the screen. Suggest how the shadow will change.

_____

_____ (2)

(c) The screen allows some of the light to pass through. What word is used to describe materials with this property?

_____ (1)

10 Katie and Finn are investigating the changes in mass of a cloth as it dries.

The children take a cloth and make it wet.

They weigh the wet cloth.

They hang the cloth in a sunny place and they weigh it at 10 minute intervals until it is dry.

They record their results in a table.

Time, in minutes	Mass of cloth, in g
0	33
10	26
20	20
30	15
40	12
50	12
60	12

(a) Plot the data from the table on the graph. (4)

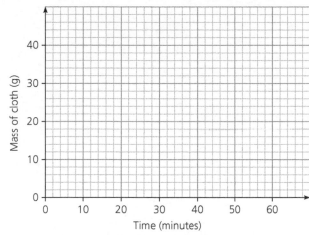

(b) Draw a smooth curve through the points. (1)

(c) Use your graph to find the mass of the cloth after 15 minutes. Show your working on the graph. (2)

(d) What is the mass of the dry cloth?

_____ (1)

Turn over to the next page.

Katie and Finn then repeat their experiment but this time they hang the cloth in a shady place.

(e) How would you expect the results for this cloth to differ from those in the first experiment? Explain your answer.

_____

_____

_____ (3)

Record your results.

Score ☐ / 80   Time ☐:☐

Answers

Paper 1: Biology (page 9)

1.   (a) **a: carbon dioxide** A plant needs carbon dioxide (and water) to make food by photosynthesis. (1)
    (b) **d: producer** The plant produces the food for the rest of the food chain. (1)
    (c) **c: hunts for food during the night** Nocturnal means 'active at night'. (1)
    (d) **c: snake** A snake is a vertebrate because it has a backbone. (1)
    (e) **c: pollination** The wind blows the long, dangling stamen and pollen is carried away to other flowers. (1)
    (f) **c: scurvy** A disease caused by lack of vitamin C. (1)

2.

	Name of part	What the part does
A	petal	brightly coloured and/or scented to attract insects for pollination (1)
B	stamen	contains pollen, which sticks to visiting insects (1)
C	carpel	the top of the carpel is sticky and the pollen from visiting insects will stick to it or it contains eggs (1)

    (b) The cells of the leaf and stem contain a green pigment called **chlorophyll**, which absorbs light energy from the **Sun** in a process called **photosynthesis**. 1 mark for each word in bold. (3)

3.   (a)

     ½ mark for each matching pair. (3)
    (b) (i) Any one of: vegetables, fruit, unprocessed cereals such as brown rice or wholemeal flour (as contained in wholemeal bread). (1)
       (ii) Fibre keeps the food moving through the intestines properly. (1)
          It adds indigestible bulk to food so that the muscles of the large intestine have something to work on. (1)

4.   (a) In a food chain the energy originally comes from the **Sun**. (1)
    (b) Breeding programmes in zoos are important to conserve **endangered** species. (1)
    (c) When an insect hatches the emerging young form is usually called a **larva**. (1)
    (d) Using a cutting to grow a new plant is an example of **asexual** reproduction. (1)
    (e) A tortoise becoming dormant in the winter is an example of **hibernation**. (1)
    (f) A camel having large feet to stop it sinking into the sand is an example of **adaptation**. (1)

5.   (a) (i) The plant is a: **producer**. (1)
      (ii) The mouse is a: **consumer** and a **herbivore**. (1)
         It is the **prey** of the owl. (1)
      (iii) The owl is a: **consumer** and a **carnivore**. (1)
         It is the **predator** of mice. (1)

(b) The arrows in a food chain represents the transfer of the energy content of food from one organism to another. (1)
Accept: The arrow means 'eaten by'.

Paper 2: Biology (page 13)

1 (a) The leaf uses **carbon dioxide** from the air and **water** from the soil to make food for the plant. (2)
   (b) Chlorophyll (1)
   (c) Capturing light energy (1)
   (d) 2 marks for either answer
       Sophie: more light means more photosynthesis and more growth.
       Ann: less light means bigger leaves to capture the limited light in shady areas. (2)
   (e) The mean is a more reliable result. (1)
       If you just compare two leaves you could select a small leaf from where the leaves are generally larger and vice versa. (1)
   (f)

       (i) vertical axis correctly labelled. (1)
       (ii) bars drawn correctly (1), neatly. (1) (2)
   (g) Label line pointing to spine on cactus. (1)

2 (a) The chimpanzees show the living processes of **nutrition**, **movement**, **growth** and **reproduction** while the toy chimpanzee does not. Other choices could include **respiration**, **excretion** and **sensitivity**.
       Any four of the living processes in bold and answer should include a comparison with the toy. (4)
   (b) Observation and recording. (2)
   (c) Chimpanzees use tools. (1)
       Chimpanzees eat meat. (1)
       Chimpanzees display emotion. (1)
   (d) Using tools enables the chimpanzees to eat more of the food available, improving their diet
       or
       Eating meat means that the chimpanzees have more protein for growth and repair in their diet
       or
       Display of emotion is communication that allows the chimpanzees to respond to danger, etc.
       2 marks for any sensible answer well explained. (2)

3 (a) Maximum of 3 marks. Deduct ½ mark for each incorrect answer.

	Seed
A	sycamore
B	dandelion
C	blackberry
D	cleaver
E	acorn
F	gorse
G	poppy

(3)

   (b) The food chain shows that squirrels eat acorns (1) but they also hide some for later (1). Some of these acorns are forgotten by the squirrels and will germinate later. (1) (3)
   (c) The coconut is dispersed by water. (1) The fibrous outer case means that it can float (1) until it reaches land, where it will germinate into a new plant. (1) (3)
   (d) Successful germination of new plants is more likely if there is less competition for **space to grow**, **light** and **water** (1 mark for idea of competing for resources and 1 mark for indicating at least one of these resources). (2)

4 (a) Animals with internal skeletons are called **vertebrates**. The skeleton is important in providing **support**, **protection** and **movement**. ½ mark for each answer. (2)

(b)

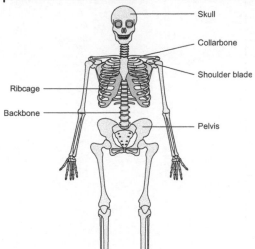

½ mark for each correct label. (3)

(c) The caterpillar grips the surface with its front end. (1)
The back end then 'walks' towards the front end and the middle bit is arched in the air. (1)
The back end is lowered to the ground and the front end moves forwards. (1)

5 Accept valid alternatives for consequences to the environment.

Human action	+ or − effect	Consequence for the environment
putting up bird boxes	+	provides places for birds to nest in areas where their normal habitat has been destroyed
keeping endangered species in zoos	+ (1)	zoos have breeding programmes to help prevent endangered species becoming extinct (1)
cutting down forests	− (1)	cutting down forests destroys the habitats of many organisms (1)
burning fossil fuels	− (1)	burning fossil fuels causes pollution which can destroy habitats (1)
creating National Parks	+ (1)	creating National Parks creates protected spaces where wildlife can survive (1)

Paper 3: Chemistry (page 20)

1 (a) **d: sieving** The larger particles stay in the sieve and the particles that are smaller than the holes of the sieve fall through. (1)

(b) **c: evaporating** Liquid changing to a gas is evaporation. (1) (Note that boiling only describes a liquid changing to a gas at its boiling point.)

(c) **d: solvent** The water is the solvent – the liquid in which the sugar (solute) is dissolved. (1)

(d) **a: coal** Coal is a solid and a fossil fuel. (1)

(e) **c: nylon** The man-made material is nylon, which is made from oil. (1) (The others all occur naturally although they are usually processed before we can use them.)

(f) **c: 36–37°C** The temperature for a healthy human is 36–37 °C. (This is the temperature at which the body works best.) (1)

2 (a)

	Does it have grains or crystals?	Is the rock igneous or sedimentary?
A	crystals (1)	igneous (1)
B	grains (1)	sedimentary (1)

(b) (i) Fossils are formed when things that have lived are trapped in **sedimentary** rock. (Note: animals, usually living in marine environments, die, fall to the sea floor and are buried in sediment.) (1)

(ii) Usually only the **hard** parts of organisms are preserved. (Note: the hard parts of an animal are either preserved intact or form a cast inside the sediment around it while the soft parts decompose.) (1)

3 (a) ½ mark for each line (2)

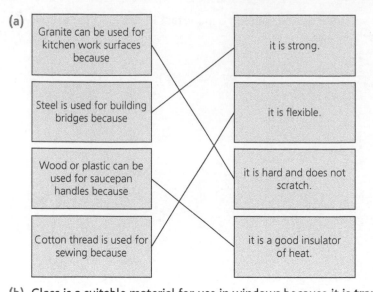

(b) Glass is a suitable material for use in windows because it is **transparent** and **waterproof**. (2)
(c) The glass is **harder** than the pencil lead. (1)
    The pencil lead is **softer** than the metal. (1)

4 (a) **Physical change** An example of melting. (1)
   (b) **Chemical change** Concrete is a new material. (1) (This is an irreversible chemical change since you cannot change the concrete back into cement, sand and gravel.)
   (c) **Chemical change** The flame on a candle is an example of burning, a chemical change. (1)
   (d) **Physical change** The wax that melts and drips down the sides of the candle is an example of a physical change. (1)
   (e) **Chemical change** A strawberry turning from green to red when it ripens is an example of a naturally occurring chemical change. (1)
   (f) **Physical change** Paint powder mixed with water is an example of a physical change. (If you leave the liquid paint and let the water evaporate you will end up with a solid lump of paint, which can be ground up into powder again.) (1)

5 (a)

Part of house	Suggestions
roof	add thick layer of roof or loft insulation material (1)
walls	cavity wall insulation (1)
windows	have thick, lined curtains/double glazing (1)
door	have curtain pulled behind door, draught proofing/double glazing in any glass panels (1)
floor	thick carpets on floorboards with a space/cavity below

(b) All of the suggestions keep the house warm by using **trapped air**, which is a **good insulator**. (1 mark for each of the phrases in bold.)

Paper 4: Chemistry (page 24)
1 (a) flexibility (1), strength (1) (2)
   (b) (i) Any sensible prediction – for example 'the flat laces will stay done up longer'. (1)
       (ii) The flat laces are less slippery/have greater friction than the round laces. (1) Answer must match prediction in part (i).
   (c) (i) Independent variable: type of shoelace. (1)
           Dependent variable: time to come undone. (1)
       (ii) Any three of:
           same length of shoelace
           same material (cotton)
           the same person should test the shoelaces each time
           the shoelaces should be in the same shoes
           the knot used to tie the laces should be the same
           the activity when testing the shoelaces should be the same. (3)
       (iii) Repeat the test (1) and average results. (1) (2)
   (d) One sensible suggestion such as:
       they could see if it depended on who was using the laces
       they could test different types of knot
       they could test shoelaces made of different materials

they could test different lengths of shoelace
they could test different brands of shoelace. (1)

2 (a) (i) Points correctly plotted (2) (deduct ½ mark for each error) (2)
    (ii) 20–50 °C straight line (1); 50–60 °C line is less steep (1) (2)

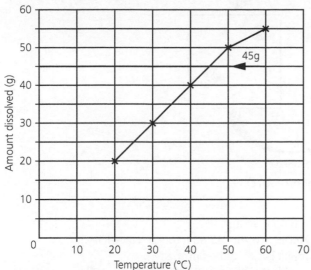

(b) (i) 45 g (1) See graph. (If graph is incorrect, value must be consistent with points plotted.)
    (ii) Any value between 58 g and 65 g. (1) Value should be consistent with graph plotted.
(c) 37 °C is the temperature of a healthy human. (1) Any less and the bath will feel cool (1)
(d) Yes (no mark)
    The graph shows that at 37 °C, 37 g of bath salts will dissolve in 1 l of water (1) so it will easily dissolve
    in 80 l of water. (1) (2)
(e) The gritty bits in Asha's bath are **insoluble** (1) solids and could be removed from the mixture in Asha's jug
    by **filtering (filtration)**. (1) (2)

3 (a) experiments (1); taking precise measurements (1)
(b) oxygen (1)
(c) 150 g (water) + 15 g (salt) = 165 g (1)
(d) hydrogen (1) and oxygen (1) (2)
(e) Burning is an example of **non-reversible** change. A material burned to produce heat is known
    as a **fuel**. A material produced both in burning and breathing is **carbon dioxide**. (3)

4 (a) (i) 42 °C (1) (units must be included).
    (ii) Thermometer correctly labelled at −7 °C. (1)

(b) Any two sensible suggestions. (2) Could include: pipes can burst, rocks crack, ice floats.
(c) No (1)
(d) Answer should include three of the points below. (3)
    Seawater is salt solution.
    Salt solution freezes at −9 °C.
    The temperature, −20 °C, was much lower than this.
    Therefore the salt solution froze.
(e) (i) B (1), D (1) (2)
    (ii) A (1) Substance A damages concrete. (1) (2)

5 (a) boxes filled correctly (see diagram below)   (2)

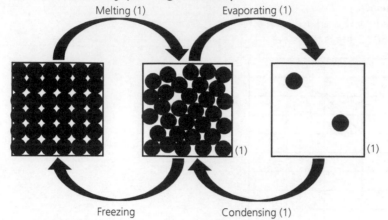

Melting (1)   Evaporating (1)

(1)   (1)

Freezing   Condensing (1)

   (b) Changes of state labelled correctly (see diagram above).   (3)

Paper 5: Physics (page 31)

1 (a) **b: a mirror** The mirror reflects light from a luminous source.   (1)
   (b) **D:** This is the metal contact with the wire. (Metals are good conductors of electricity.) (1)
   (c) **c: upthrust** The upward force on a floating object is called the upthrust.   (1)
   (d) **d: Diagram D** A smaller force is able to lift an object with a greater force using a pulley. (1)
   (e) **a: higher pitched** Increasing the rate of vibrations produces a higher pitched sound. (1)
   (f) **b: galaxy** The Milky Way is a galaxy. Our solar system is part of the Milky Way. (1)

2

	Useful friction	Nuisance friction	Explanation
A	✓		Friction between the wheels and the road is called grip; it prevents the bike skidding and allows the cyclist to control the motion of the bike.
B		✓ (1)	Friction between the air and the cyclist is called air resistance and this slows the cyclist down. (1)
C	✓ (1)		Friction between the wheel rim and the brake blocks when braking is important as it allows the cyclist to stop quickly. (1)
D		✓ (1)	Friction between the moving parts of the chain and cogs turning the wheels is called contact friction and it makes it more difficult for the cyclist to pedal and turn the wheels. (1)

3 (a) (i)   The north-seeking pole of a magnet will **repel** another north-seeking pole.   (1)
        (ii) The south-seeking pole of a magnet will **attract** an iron bar.   (1)
   (b)

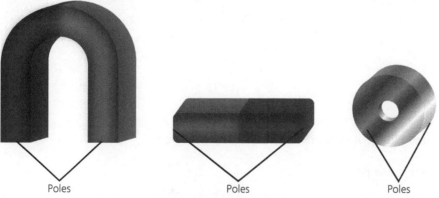

Poles   Poles   Poles

Point to the two flat faces.   Point to the left-hand and   Point to the front and back
                               right-hand ends.   circular faces.   (3)

        1 mark for identifying the poles correctly on each magnet.
   (c) A compass needle aligning in a north-south direction in the Earth's magnetic field.   (1)
        Accept valid alternatives.

4 (a) The roughly circular path that the Earth takes around the Sun is known as its **orbit**. (1)
   (b) Day and night are caused by the spin of the Earth on its own axis.   (1)
        The half facing the Sun is in daylight and the half facing away is in night-time.   (1)

Science Practice Papers published by Galore Park

(c) It takes approximately **27 days** for the Moon to go around the Earth (accept 28 days). (1)
(d) Ships disappearing over the horizon. (1)
Photographs taken from orbiting spacecraft, have revealed that the Earth is spherical. (1)

5 (a)

Correct symbols for **two cells**, **an open switch** and a **buzzer** connected in **series**. (4)
(b) Two of: one or more of the cells could have run down; the buzzer could be broken. (2)

Paper 6: Physics (page 35)

1 (a) (Electrical) insulators (1)
   (b) (i) The metal materials are conductors of electricity. (1)
      (ii) The inside part of the green colouring pencil does not conduct electricity. (1)
         The inside part of the 2B pencil conducts electricity poorly. (1)
      (iii) That metals conduct electricity but **most** non-metals do not. (1)
   (c) (i) Only the metal in each wire should be different. (1)
         The wires should all be the same length (1) and the same diameter (1).
      (ii) That some metals are better conductors than others. (1)
      (iii) copper (1)

2 (a)

Seen as a luminous source	Seen by reflected light
computer screensaver (½)	black tinted window (½)
television picture (½)	mirror (½)
the Sun (½)	radio (½)

   (b) The light ray should be drawn as a straight line from the light to the girl's eyes (1) with an arrow pointing from the light to the girl. (1) (2)
   (c) damages eyes (1)
   (d) (i) Two straight lines drawn from A and B to the screen, just touching top and bottom of the shadow puppet (see image below). (1)

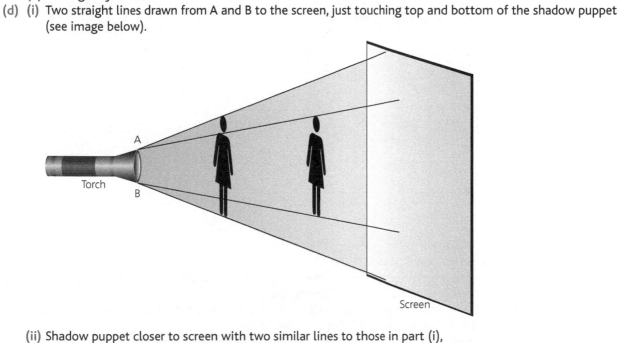

      (ii) Shadow puppet closer to screen with two similar lines to those in part (i), now indicating a shorter shadow. (1)
      (iii) The closer the shadow puppet is to the screen (1) the smaller the shadow (1) or vice versa. Answer should be a logically constructed sentence. (2)

3 (a) Sounds are made when objects vibrate. Increasing the size of the vibration makes the sound **louder**. Increasing the speed of vibration makes the sound **higher** pitched.
      When playing a guitar, shortening the string increases the **pitch** and plucking it gently will make a **quieter** sound.
      A thick string has a **lower** pitched sound than a thin string. Tightening the string will make the note **higher**. (6)
   (b) (i) The earlobe collects sound and funnels it into the ear. (1)
      (ii) The eardrum (1)

4 (a) force (1)
   (b) Gravity is a force of attraction (1) between two objects.
      Accept valid alternatives, e.g. keeps planets in orbit around the Sun. (2)
   (c) Gravitational **force** (1) keeps the Moon in **orbit** (1) around the **Earth** (1) and the **planets** (1) on their paths around the **Sun** (1). (5)
   (d)

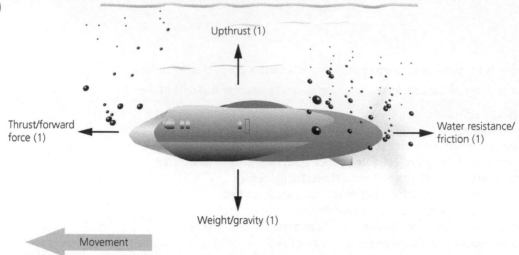

Upthrust (1)

Thrust/forward force (1)

Water resistance/ friction (1)

Weight/gravity (1)

Movement

5 (a) (i) Scales to make full use of grid (1) correctly numbered. (1) (2)
      (ii) Points correctly plotted. (½ mark for each point, 3 marks in total.) (3)
      (iii) Straight line drawn to best match the points. (1)
   (b) friction (1)
   (c) No. (1)
      If one dog can pull up to 300 N, four dogs together can only pull up to 1200 N. (2)

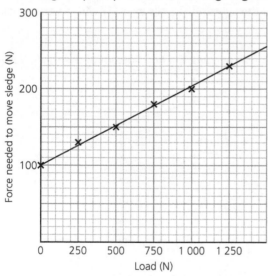

Paper 7: 11+ Practice Paper (page 42)
1 (a) **d: protein** Fish is a good source of protein. (1)
   (b) **b: air, water and warmth** are needed for germination. (1) (Remember: WOW – Water, Oxygen (air), Warmth)
   (c) **c: translucent** A translucent material lets some light through it. (1)
   (d) **a: a vacuum** Sound cannot travel through a vacuum. (1) (There are no particles in a vacuum to vibrate.)
   (e) **b: solute** The material that is dissolved in a solution is the solute. (1)
   (f) **b: copper** Copper is a thermal conductor. (1) Other options are all thermal insulators.
   (g) **c: fur** Fur is a diagnostic feature of mammals in classification. (1)
   (h) **d: Sun** The Sun is the star around which the Earth orbits (1)
   (i) **d: migration** When an animal moves long distances for food/to breed. (1)
   (j) **c: menstruation starting** A girl's ovaries start to release eggs as part of a monthly or menstrual cycle. (1)
2 (a) A: Incisor (1)
      B: Pre-molar (1)
      C: Canine (1)
      D: Molar (1)

(b) Any two suggestions from the following suggestions well explained (2 marks each, maximum 4 marks).
Human: smaller canines, less meat in diet than dog
Human: teeth are all similar in size, more varied diet than dog (4)
Dog: has large molars for crunching bones in its diet while humans have smaller molars for chewing foods in their diets. (4)

(c) Brush teeth regularly; floss between teeth; use fluoride toothpaste; use mouthwash; visit dentist regularly. Any three valid suggestions. (3)

3 (a) drum: skin vibrates (1)
pan pipes: air in pipes vibrates (1)
guitar: strings vibrate (1)

(b) Hit the drum harder to make it louder. (1)

(c) Shorten string by pressing against a fret (1) or play a thinner string. (1) (2)

(d) The sound would get fainter. (1)

(e) Sound travels from the 'one-man band' through the air as air vibrations. (1)
The air vibrations then hit the eardrum and it vibrates. (1)
The eardrum vibrations send messages (via the inner ear and nerves) to the brain which are interpreted as sound. (1)

(f) The audible range becomes less as you get older (1) with higher notes being more difficult to hear. (1) (2)

4 (a) A: evaporation (1)
B: condensation (1)

(b) (i) The Sun (1)
(ii) Wind (wind can give energy to particles at the surface of a liquid causing them to escape as a gas). (1)

(c) A: freezing (1)
B: melting (1)

(d) In the diagrams the:
solid particles should be regularly arranged; particles touching (1)
liquid particles should be irregularly arranged; particles mostly touching (1)
gas particles should be few in number (about three); particles widely spaced. (1)

5 (a)

(i) Suitable scale for horizontal axis. (1)
(ii) Data correctly plotted, but allow reasonable approximation for first two points to allow for difficult scale (deduct ½ mark for each incorrect plot). (3)
(iii) A smooth curve through the data points. (1)

(b) Baby has a mass that is too small to be measured (1) or mother does not know she is pregnant so it is not possible to measure the mass (1).
Accept any one sensible answer. (1)

(c) 0 to 20 weeks: growth is very slow as shown by the flatter slope of the graph.
20 to 40 weeks: growth is very rapid as shown by the steeper slope of the graph. (1)

(d) Two suggestions: By 20 weeks the baby has all of its limbs (1); has all of its organs (1); it is a very small, fully developed baby. (1) (2)
Accept any two sensible suggestions using information from the scan.

(e) The data in the table is an average and does not represent individual cases (which could depend upon different nutrition levels of the mother, health of the mother, genetic differences, etc.). (1)

6 (a) Copper is used for the wires in an electrical circuit because it is a good **electrical conductor**. (1)

(b) Plastic is used for the outside of a plug because it is a good **electrical insulator**. (1)

(c) When there is too much current in a circuit the fuse will **melt**. (1)

(d) A non-metal that conducts electricity is **graphite**. (1)

(e) The components in the above circuit are connected in **series**. (1)

(f) If you add an extra cell to the circuit in part (e) the bulb will shine **brightly**. (1)

(g) The components in the circuit in part (e) are a cell, a bulb and a **closed switch**. (1)

7 (a) It is important to record results to avoid replicating experiments that did not work (1) also so that possibly useful results are available later. (1) (2)
Accept any sensible answer.

(b) The sticky note could be removed (1) without leaving a mark on the page. (1) (2)

(c) A sticky board would attract dust or dirt which would stop it working effectively. (1)

(d) Use the glue to stick the top of the plastic strip to the edge of the table.
Allow glue to dry.
Attach the hook of the weight holder to plastic strip using the hole at the bottom of plastic strip.
Add weights one at a time until glue fails.
Repeat experiment using different types of glue.
Record results.
Accept any sensible method with at least four instructions written in a logical order. (4)

(e) Only change the type of glue (independent variable); (1)
all other variables kept the same (control variables). (Any two of: amount of glue, size of plastic strip, test set up, time and place of investigation.) (2)

8

Substance	Acid, alkaline or neutral	Blue litmus paper	Pink litmus paper
lemon juice	acid	turns pink (1)	no change/stays pink (1)
salt water	neutral	no change/stays blue (1)	no change/stays pink (1)
sink cleaner	alkaline	no change/stays blue (1)	turns blue (1)
white vinegar	acid	turns pink (1)	no change/stays pink (1)

(Remember: Blue litmus paper turns pink in acid solutions. Pink litmus paper turns blue in alkaline solutions. Both colours of litmus paper stay the same in neutral solutions.)

Paper 8: 11+ Practice Paper (page 50)

1 (a) **b: movement** Plants move very slowly to face the Sun but the movement is difficult to observe. (1)

(b) **d: stem** The stem contains vessels that carry water to all parts of the plant. (1)

(c) **b: an earthworm** An earthworm does not have a backbone. (1)

(d) **d: temporary** A perforated eardrum will usually heal. (It will be painful and there will be hearing loss until it gets better.) (1)

(e) **d: sieving** The salt grains are smaller than the rice and will pass through a sieve. (1)

(f) **a: 6:00 a.m.** Your shadow will be longest when the Sun is low in the sky. (1)

(g) **c: liquid to solid** The water freezes to become ice. (1)

(h) **a: insoluble** Chalk does not dissolve in water. (1)

(i) **a: crystals** Volcanic rocks have large crystals when the molten rock cools slowly. (1)

(j) **d: reflected** Smooth, shiny surfaces such as mirrors reflect light. (1)

2 (a)

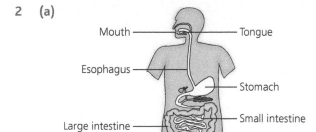

1 mark for each correct label (6)

(b)

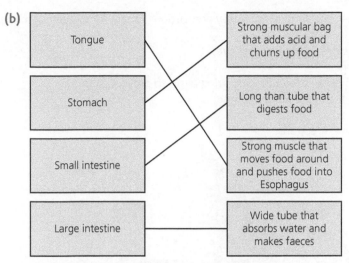

      1 mark for each correct line joining the boxes. (4)

3 **(a)** Neat diagram **(1)** Correct diagrams and labels for filter funnel **(1)**, filter paper **(1)** and beaker **(1)** as below (clamp stand need not be shown).

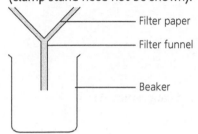

 (4)

    (b) **(i)** filtrate (1)
        (ii) residue (1)
    (c) The water was not pure. (1)
    (d) There were soluble **(1)** substances dissolved **(1)** in the water.
    (e) The toxic materials would kill animals in the stream. **(1)** Accept any valid answer.

4 **(a)** Only change the type of paper (independent variable). (1)
        Measure how far the plane travels (dependent variable). (1)
        Keep all other variables the same (control variables) **(1)**: the design of the plane; the force used to
        throw it. **(2)** Accept other valid control variables.
    (b) **(i)** To make sure that results are reliable. (1)
        (ii) 4.0 (written in table) (1)
    (c) **(i)** Vertical axis labelled 'Distance thrown, in m'. (1)
        (ii) Three bars drawn neatly **(1)** and accurately **(1)**. Accept bar drawn to show the answer for 'card' given in (b) (ii).

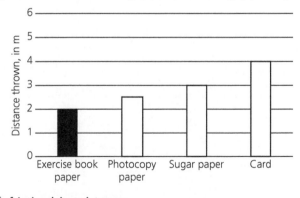

    (d) **(i)** friction/air resistance (1)
        (ii) The card is the smoothest material and has least friction or card is the stiffest material so travels smoothly
        with less friction. **(1)** Accept valid alternatives.

5 **(a)** When the salt solution is heated the liquid **evaporates** **(1)** to become a **gas**. **(1)** (2)
    (b) Salt crystals will spit out of the evaporating basin. (1)
    (c) The beaker is good for larger volumes of solution but will break if it is heated when the mixture becomes
        drier, **(1)** the evaporating basin can be heated to higher temperatures. **(1)** (2)

(d) Two safety precautions explained for 2 marks each. 1 mark for safety precaution; 1 mark for explanation. These could include:
Wear safety glasses to prevent hot water or salt spitting into her eyes.
Avoid touching hot apparatus to prevent burns.
Use heatproof gloves when transferring hot salt solution to evaporating basin to prevent burns.
Keep hair tied back so that it does not catch fire in the Bunsen flame. (4)
Accept other valid suggestions.

6 (a) (i) Two arrows of equal length and in opposite directions.

(2)

(ii) No (1) The forces are of equal strength (1) in opposite directions and cancel out. (1) (3)
(iii) One end should push (1) and the other end should pull. (1) (2)

(b)

Example	Push	Pull	Both
pedalling a bike	✓ (½)		
opening a box		✓ (½)	
spinning a coin			✓ (½)
shutting the curtains		✓ (½)	
closing a cupboard door	✓ (½)		
playing on a swing			✓ (½)

(3)

7 (a) The names of the animals and plants were too long to remember. (1) Accept valid alternatives.
(b) Fungi do not have chlorophyll and feed by breaking down organic matter. (1)
(c)

```
                        Living things
        ┌──────────┬──────────┬──────────┬──────────┐
     Animals     Plants    Bacteria     Fungi   Single-celled
        │                                          organisms
   ┌────┴────┐
Vertebrates  Invertebrates
```

(3)

The five Kingdoms (in the second row) can be in any order except that 'animals' that should link to the third row of the chart. Deduct 1 mark for each error.

(d)

Vertebrate group	Type of skin
mammals (1)	hair/fur
birds	feathers (1)
fish (1)	wet scales
amphibians (1)	moist skin
reptiles	dry scales (1)

8 (a) A candle **burning** as a luminous source. (1)
A bicycle **rusting** when left out in the rain. (1)
Mixing sand, cement and water which **sets** to form concrete. (1)
Mixing eggs, sugar, butter and flour and **baking** to make a cake. (1)
Strawberries **ripening** and changing colour from green to red. (1)

(b) (i) Burning fossil fuels and producing polluting gases and soot/Farmers adding pesticides that also kill other animals. Any sensible suggestion for 1 mark and explained for second mark. (2)

(ii) Dropping litter/rubbish dumps. (1) Any sensible suggestion for 1 mark. (1)

(iii) Volcanoes erupting, forest fires, tsunamis, most natural disasters. Any sensible suggestion for 1 mark. (1)

Paper 9: 11+ Practice Paper (page 58)

1 (a)

	Organ
A	brain (1)
B	lung(s) (1)
C	heart (1)
D	stomach (1)
E	(small) intestine (1)

(b) The skeleton supports the soft parts of the body, (1) protects the delicate organs (1) and helps the body to move (1)

(c) Organ A is inside the **skull** (1) and organ B is inside the **ribs/ribcage.** (1) (2)

2 (a) Pink litmus paper will stay pink (1) and blue litmus paper will stay blue (1) in a neutral solution.

(b) indicator (1)

(c) Clear colour changes are more likely in brightly coloured pigments. (1) Any sensible suggestion.

(d) The pigments in the three extracts are all similar. (1)

(e) Three points, from:
• Red wine contains a pigment that is an indicator.
• The pigment in the wine is red before alkali is added suggesting that red wine is an acid.
• The pigment in the wine is green in alkaline solutions.
• The pigments in the wine show a similar change in colour to other plant extracts tested.
Accept valid alternatives. (3)

3 In each of the answers below, the parts in bold may be worded differently but should reflect the answers as stated.

(a) The bulb in circuit **B** is **brighter** (1) compared with the bulb in circuit **A** because **there are more cells in the circuit.** (1) (2)

(b) The motor in circuit **B** turns **in the opposite direction** (1) compared with the motor in circuit **A** because the **cells are facing in the opposite direction.** (1) (2)

(c) The bulb in circuit **B** is **off/not working** (1) compared with the bulb in circuit A because **the cells are facing each other.** (1) (2)

(d) The buzzer in circuit **B** is **off/not working** (1) compared with the bulb in circuit A because **the switch is open/there is not a complete circuit.** (1) (2)

(e) The bulb in circuit **B** is **the same brightness** (1) compared with the bulb in circuit A because **although there are three cells in circuit A, two are facing each other and cancel out, so there is one cell's worth of current flowing round each circuit.** (1) (2)

4 (a) (i) Do thicker layers of insulating material reduce heat loss from the water? (1)
Accept valid alternatives but answer must refer to the thickness of the insulating material (the independent variable) and the heat loss (the dependent variable).

(ii) starting temperature of water, volume of water, size and shape of beaker (3)

(iii) Less heat loss from the surface of the water. (1)

(b) 25 °C (1) – the temperature will not fall below the temperature of the surroundings/room temperature. (1)

(c) (i) points correctly plotted (2) Deduct ½ mark for each error. (2)
  (ii) smooth, neat curve joining points (1)

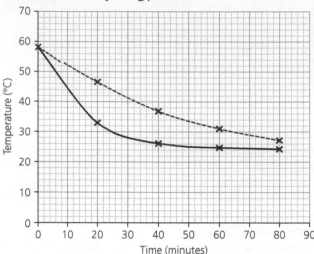

(d) smooth, shallow curve (1) above (1) the line for 2 cm thickness of cotton wool.

5 (a) Sexual reproduction in plants and animals involves the **fusion** (1) of male and female sex cells in a process called **fertilisation.** (1) In plants, the male sex cells are contained in the **stamen** (1) and transferred to the female sex cells in the **carpel** (1) in a process called pollination. Sexual reproduction produces offspring with characteristics that vary from the characteristics of the parent animal or plant.
Some microscopic animals reproduce asexually by splitting into two when they reach a certain size. Plants can reproduce asexually by a variety of methods. The offspring are **identical** (1) to the parent plant. Plants often use both methods of reproduction.

(b)

Description	Sexual	Asexual	Both
A strawberry plant has small white flowers and has runners which grow into new plants			✓ (1)
An amoeba consists of one cell which divides into two when it gets large		✓ (1)	
An oak tree uses the wind for pollination and squirrels for seed dispersal	✓ (1)		
A pair of rabbits could produce up to 96 baby rabbits in a year	✓ (1)		
A hydra is a tiny freshwater animal about 1 cm long. It reproduces by budding where the offspring grow out of the body of the parent		✓ (1)	

6 (a) Midday/12 o'clock (1) the Sun is directly overhead. (1) (2)
  (b) Morning (1) the Sun rises in the east (1) and the shadow will point west. (1) (3)
  (c) It is night-time. (1)
  (d) (i) The Earth is spinning on its own axis. (1)
    (ii) The Earth is orbiting the Sun. (1)
    (iii) The Moon is orbiting the Earth. (1)

7 (a)

Field notebook entry	Rock
black rock that needed a hand lens to show that it was made of tiny crystals	basalt (1)
yellow/brown rock that was made up of clearly visible grains	sandstone (1)
grey rock that needed a hand lens to show that it was made of tiny grains	mudstone (1)
grey rock that was made up of clearly visible large crystals	granite (1)

(4)

(b) colour (1)
(c) (i) Size of grain as in 'large' is too vague a term. The large grains in one rock may seem small when looked at alongside another rock with even larger grains. Accept sensible suggestions well explained. (2)
  (ii) 'Does the rock have fossils?' (1)
(d) limestone (1)

8 (a)

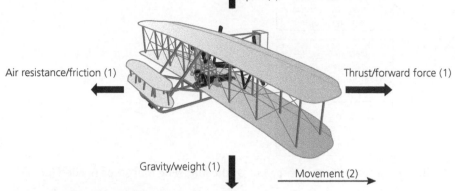

Uplift (1)

Air resistance/friction (1)

Thrust/forward force (1)

Gravity/weight (1)

Movement (2)

1 mark for each labelled arrow. (4)

(b) (i) Arrow pointing to right (1), labelled movement (1) as in diagram above.

(ii) The force arrow pointing to the right is longer than the force arrow pointing to the left. (1)

(c) They observed the wings of birds. (1)

9 (a) The arrows represent the transfer of energy between organisms in the food chain/accept 'they mean eaten by'. (1)

(b) (i) The Sun. (1)

(ii) Chlorophyll in plants captures energy from the Sun (1) as they make their own food in the process called photosynthesis. (1) (2)

Paper 10: 11+ Practice Paper (page 67)

1 (a)

Example	Nutrition	Movement	Growth	Reproduction
a lion chasing a gazelle		✓ (1)		
green leaves on a plant absorbing energy from the Sun	✓ (1)			
digested foods in an animal being used to build new cells			✓ (1)	
a flower producing seeds which then germinate to produce new plants			✓ (1)	✓ (1)
a rabbit running away from a predator hawk		✓ (1)		

(6)

(b) Leaves turn to the Sun to absorb more sunlight (1) for photosynthesis (1) (2)

(c) The child can grow (1) and eventually reproduce (1) while the robot does not exhibit these life processes. The child can feed (1) but the robot cannot. Any two sensible suggestions, each of which should give a comparison between the child and the robot. (2)

2 (a) (i) Straight line from Sun to Moon (1) with arrow pointing to Moon (1) (see diagram below).

(ii) Straight line from Moon to A on Earth (1) with arrow pointing to Earth (1) (see diagram below).

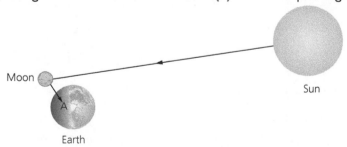

Moon

Sun

A

Earth

(b) The Sun is a luminous (1) object because it gives out light. (1) The Moon is a non-luminous (1) object and can only be seen by reflected (1) light. On Earth we can see the Sun and the Moon because the atmosphere is transparent. (1) It is dangerous to look directly at the Sun. (1) (6)

3 (a)

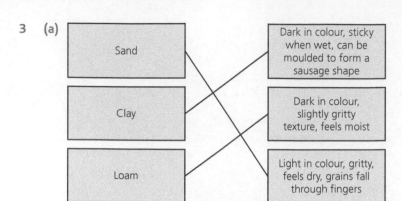

(3)

(b) Method should be clear and follow a logical sequence. Three out of the following four steps needed for 3 marks.

Fill the beaker with equal amounts of water for each test.

Start clock and pour water onto soil sample.

After one minute (or specified amount of time less than 5 minutes) see how much water is in the measuring cylinder.

Record results. (3)

(c) (i) Sand (1)

(ii) Sand has large particles with spaces between for water to drain through/no organic matter to absorb and retain water. One suggestion (1) well explained (1).

(d) (i) Humus (1)

(ii) Some water retention in soil but not so much that the soil is waterlogged/contains organic matter, which enriches the soil providing minerals for healthy plant growth. One suggestion (1) well explained (1).

4 (a) Pesticides are used to kill insects, weeds and small animals (1) that eat crops. (1) (2)

(b) Organic farming does not use pesticides so harmless organisms are not killed/Organic farming is better for the environment. (1)

People do not get sick through eating harmful chemicals used as pesticides. (1)

(However, pesticides are now tested to make sure that they do not affect people's health before they are allowed for use.)

(c) Thin shells will break when the adult sits on them in the nest and the chicks will die (1) leading to a decrease in the bird populations. (1) (2)

(d)

(2)

Deduct 1 mark for each error.

5 (a)

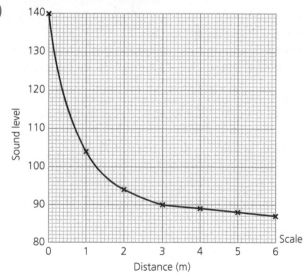

(i) vertical axis correctly labelled (1)

(ii) suitable scale on horizontal axis (1)

(iii) points correctly plotted (3).
Deduct half a mark for each error.

(iv) smooth, neat curve joining points (1)

(6)

(b) As distance from the source of sound/baby crying increased (1) the sound level decreased. (1) (2)

(c) Yes: the sound level of the baby crying is above the pain threshold (1) when Sarah is close to/holding the baby. (1) (2)

6 (a) The explanations here and in part D should refer to both of the materials in the mixture for 2 marks.

　　(i) Method: magnetism (1)

　　The method works because the iron filings are a magnetic material and attracted to the magnet (1) leaving the sand, a non-magnetic material, behind. (1) (2)

　　(ii) Method: filtration (1)

　　The method works because the chalk is an insoluble solid that will remain as a residue on the filter paper (1) while the water is a liquid and will pass through the filter paper (as the filtrate). (1) (2)

　(b) There is only 1 mark for each answer but it is important to get into the habit of showing your workings, even if they are very simple.

　　(i) 350 − 250 = 100 g (1)

　　(ii) 253 − 250 = 3 g (1)

　　(iii) 100 − 3 = 97 g (1)

　(c) 97 g (water) + 3 g (salt) = 100 g (salt/seawater) (this shows that the mass of the constituent parts of the mixture equals the mass of the mixture) (1)

　(d) The method works because the water part of the mixture is a liquid which will evaporate (1) while the salt is a solid and will not evaporate. (1) (2)

7 (a) Diagram to be drawn as below:

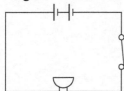

1 mark each for correct symbols for: 2 cells, a buzzer and a closed switch (3) in series (1) neatly drawn with ruled lines for connecting wires. (1) (5)

　(b) Archie short circuited the buzzer by adding a wire that 'by-passed' it. (1) (See diagram below.)

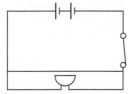

This is dangerous because the cells would continue to get hotter and could catch fire. (1) (2)

8 (a) A balanced diet contains all of the food groups (1) in the correct amounts/proportions. (1) (2)

　(b) (i) 117 g (1)

　　(ii) water (1)

　(c) Accept sensible alternatives.

	Comments
protein	The ready meal provides 32 out of the 45 g of protein that I need so I do not need much more protein today.
carbohydrate	The ready meal provides 60 out of the 320 g of carbohydrate that I need so I should try to eat some foods high in carbohydrates (but be careful not to eat too much sugar). (1)
fat	The ready meal provides 20 out of the 90 g of fat that I need so I can eat some more. (1)
fibre	The ready meal provides 4 out of the 24 g of fibre that I need so I should eat some high fibre foods as part of my other meals. (1)
salt	The ready meal provides 1 out of the 6 g of salt that I need so I can eat some more (but should be aware that some foods are particularly high in salt). (1)

(4)

　(d) Breakfast B (1), dinner/supper A (1), snacks B (1) (3)

1 (a) **d: stem** the stem transports food made in leaves to all parts of the plant. (1)
   (b) **b: make food** The roots are in the soil and receive no light so they cannot photosynthesise. (1)
   (c) **b: magnetism** Iron filings or a compass needle will be affected by the magnet, even when they are not touching it. (1)
   (d) **c: freezing** The water condenses as it cools to form water which then freezes to become snow. (1)
   (e) **c: dissolving** Can be reversed by separating the parts of the mixture (solution). (1)
   (f) **a: gear** The teeth on the outer rim engage with teeth of other gears. (1)
   (g) **d: reflect light** The Moon and the planets reflect light from the Sun. (1)
   (h) **c: three body parts** Insects are characterised by having six legs and three body parts. (1)
   (i) **c: N** N is for newton and forces are measured in newtons. (1)
   (j) **c: a once-living thing preserved in sedimentary rock** The trilobite was alive before it died and became preserved as a fossil. (1)

2 (a) Answers will depend on method of sorting suggested.
      1 mark for any sensible method and 1 mark for each column of the table correctly completed.
      Answers could include:
      Question: Is it alive or dead? (1)

group 1: alive	group 2: dead
guinea pig	sheep skull
locust	dried flowers
cactus	(1)
sunflower plant	
(1)	

(2)

Question: Is it an animal or a plant? (1)

group 1: animal	group 2: plant
guinea pig	dried flowers
locust	cactus
sheep skull	sunflower plant
(1)	(1)

(2)

Other questions could include 'Is it able to move or not?', 'Does it have legs or no legs?', etc.

   (b) (i) Labelling the vertical axis. (1)
      (ii) Plotting the missing bars. (½ mark each and 1 mark for accuracy and neatness.) (3)

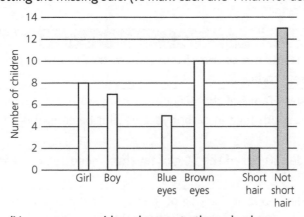

   (c) (i) Yes/No answers provide a clear route through a key. (1)
      (ii) 'Very short' is relative term/not clear what 'very short' means/hair can be cut or grow so the identification question will not produce a consistent answer (any sensible answer) (1)

3 (a) First column of table as below   (3)
   (b) Last column of table as below (3) Answers should reflect but not copy information from the paragraph.

Sieve	Particle size	Mass of soil, in g	Description of contents
A (1)	more than 2.0 mm	10	stony, contains dead wood and leaves (1)
B (1)	0.1 mm to 2.0 mm	130	medium sized particles with black material (1)
C (1)	less than 0.1 mm	60	very fine particles mixed with black material (1)

   (c) humus   (1)
   (d) clay   (1)
   (e) (i) loam   (1)
       (ii) The soil contains a mixture of sand and clay without being predominantly one or the other (1); contains
            organic material. (1)   (2)
4 (a) The metal ends of the fuse are connected to terminals as a part of a **series** (1) circuit. The electricity can pass
       through one metal end, along a thin **wire** (1) to the other metal end of the fuse.
       If too much electricity flows through the fuse the wire will get hot and **melt** (1) and this **breaks** (1) the circuit.
       Fuses are **safety** (1) devices used to protect delicate electric components and to prevent electric **shocks** (1).   (6)
   (b) Two suggestions which could include: keep water away from electricity, do not put anything other than a plug
       into an electric socket, make sure wires and plugs are properly insulated, use batteries/cells in experiments not
       mains electricity. Accept sensible alternatives.   (2)
5 (a) (i) The water freezes, expands (1) and the hosepipe will split (1)   (2)
       (ii) The water freezes, expands (1) and the stones of the ancient building will be forced apart and
            break away (1)   (2)
   (b) When the water freezes it could split the pipes (1). When the water melts the pipes will leak and cause more
       damage. (1)   (2)
   (c) (i) Flexible (1) so that the material can be easily wrapped around the pipes. (1)   (2)
            A good insulator (containing trapped air) (1) to reduce heat loss from the pipes. (1)   (2)
       (ii) Any gaps around the pipes where they go through an outside wall should be sealed to prevent heat from
            inside the house escaping.   (1)
       (iii) 10 °C – this will prevent the pipes from freezing without heating the house too much.   (1)
6 (a) (i) It was too cold for the seeds to germinate.   (1)
       (ii) Either seeds were waterlogged and rotted or plastic wrap kept out air containing oxygen.   (1)
       (iii) Seeds were not watered and it was too dry for them to germinate.   (1)
   (b) Seeds had warmth (1), oxygen (from the air) (1) and water (1) for germination (remember WOW!).   (3)
   (c) The bean does not need to be planted in soil to germinate.   (1)
       Light is not a factor that affects germination.   (1)
   (d) Points should be in correct order for 3 marks:
       root grows first   (1)
       followed by shoot   (1)
       first leaves appear, plant starts photosynthesising and grows rapidly.   (1)
       (Note: food supplies in the seed provide materials for growth until first leaves appear and photosynthesis can
       start.)
7 (a) (i) That the Earth is a sphere (1) and that planets move in roughly circular orbits. (1)   (2)
       (ii) That the other planets move around the Earth.   (1)
   (b) His ideas were opposed because Ptolemy's ideas had been accepted for almost 1400 years. Accept sensible (1)
       alternatives such as going against the views of the established church.
   (c) (i)   The Sun is an example of a **star**.   (1)
       (ii) The Milky Way is an example of a **galaxy**.   (1)
       (iii) The Sun, the planets and their moons all have a shape that is **spherical** (accept circular).   (1)
       (iv) The force that keeps the Moon in its orbit around the Earth and the Earth in its orbit around the Sun
             is **gravity**.   (1)
   (d) The Earth rotates/spins on its axis. (1) One half will be facing the Sun and it is day, the other half will be facing
       away from the Sun and it is night. (1)   (2)

8 1 mark for each statement correctly connected to its part in the investigation.

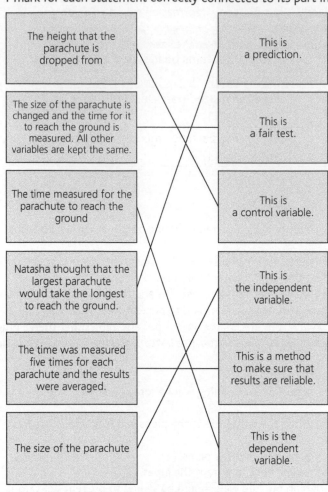

The height that the parachute is dropped from		This is a prediction.
The size of the parachute is changed and the time for it to reach the ground is measured. All other variables are kept the same.		This is a fair test.
The time measured for the parachute to reach the ground		This is a control variable.
Natasha thought that the largest parachute would take the longest to reach the ground.		This is the independent variable.
The time was measured five times for each parachute and the results were averaged.		This is a method to make sure that results are reliable.
The size of the parachute		This is the dependent variable.

(6)

Paper 12: 11+ Mock Exam (page 88)

1 (a) Sounds are made when objects **vibrate**. (1)
   (b) Copper is used for household wiring because it is an electrical **conductor**. (1)
   (c) A tadpole is a young **amphibian**. (1)
   (d) Adding another bulb to a circuit makes the bulbs **dimmer**. (1)
   (e) Sugar dissolves in water because it is **soluble**. (1)
   (f) The force that slows a spacecraft re-entering the atmosphere is **air resistance**. (1)
   (g) The remains of once-living things found in rocks are called **fossils**. (1)

2 (a) (i) reversible (1)
       (ii) non-reversible (1)
       (iii) non-reversible (1)
       (iv) reversible (1)
   (b) (i) Water (1) and air/oxygen (1) must both be present to cause rusting to occur. (2)
       (ii) The spade is made from iron/steel (1) and the greenhouse is made from a different metal that does not rust. (1)/ The greenhouse has been painted/galvanised (1) but the spade is untreated. (1) Accept valid alternatives. (4)
       (iii) Bob could cover the surface of the spade with oil. This would keep water and air from coming into contact with the surface of the metal.
       He could make sure that he cleans and dries his spade immediately after use and keeps it in a dry place. This would work by keeping water away from the surface of the metal.
       (1 mark for the method of prevention, 1 mark for the explanation of how it works.) Note: Painting and galvanising, which are also methods of rust prevention, would not be practical in this case as the paint or zinc would chip off during use. (2)

3 (a) Chimpanzees are mammals. (1) They have fur (1) and feed their young on milk (1). (3)
   (b) Chimpanzees are omnivores. (1) (They eat plant and animal material.)
   (c) Their hands and feet are shaped to help them to grip in the trees. They have learnt to use tools to help them find food. They can move around on the ground and in the trees. (1 mark each for any two valid suggestions.) (2)

(d) Endangered species have so few individuals left that they are in danger of extinction. (1)

(e) People can help by protecting the chimpanzees' habitat. (Accept valid alternatives, e.g. by captive breeding and reintroduction.) (1)

4 (a) 20 + 10 = 30, 100 − 30 = 70 g of the smallest particles. (1 mark for the correct answer with the correct unit, 1 mark for showing working.) (2)

(b) Soil A is sand. (It has mostly larger particles.)
Soil C is clay. (It has mostly the smallest sized particles.)
Soil B is loam. (It has a roughly even distribution of particle sizes.) (2)
(2 marks for all three correct, 1 mark for 1 correct)

(c) This material is called humus. (1)

(d) They could weigh equal masses (1) of the three soils and place each in turn in the apparatus. The water would be allowed to drain through for the same length of time (1) (accept a given sensible time, e.g. 1 minute). The volumes of water collected in the measuring cylinder for each sample would be compared. (1) The one with the most water allows water through quickest. (1) (Accept valid alternative methods.) (4)

(e) Soil A would drain most quickly (1) because it contains the biggest particles (1). (2)

5 Circuit B The lamp will be brighter than A. (There is an additional cell in this circuit) (1)
Circuit C The lamp will be the same as Circuit A. (Two cells facing in one direction and one in the opposite direction have the same effect as one cell). (1)
Circuit D The lamp will be off. (There is a short circuit so the electricity does not flow through the lamp.) (1)

6 (a) Plants need light, water and warmth to grow well (1 mark each). (Accept mineral salts.) (3)

(b) Plants use the process called photosynthesis (1) to make their own food.

(c) Compost (dead plant material, i.e. humus) adds nutrients (mineral salts) to the soil/helps to retain water/improves soil structure. (1 mark each for any two suggestions.) (2)

(d) He could shake the soil with water (1) filter the mixture (1) and test the filtrate with litmus/red cabbage extract. (1) If the soil is alkaline, the litmus will turn blue/red cabbage extract will turn green (1). (4)

7 (a) Thermometer A reads 31 °C. (1)
Thermometer B reads −5 °C. (1)
Thermometer C reads 80 °C. (1)

(b) (i) The force is measured in newtons (N). (1)
(ii) The distance is measured in kilometres (km). (1)
(iii) The volume is measured in cubic centimetres (cm^3). (1)
(iv) Your mass is measured in kilograms (kg). (1)
(v) The area is measured in square metres (m^2). (1)

(c) Measure the thickness of a known number of pages (1) and then divide by the number of pages to give the thickness of one page. (1)

8 (a) The two planets with orbits closest to Pluto are Neptune and Uranus. (1 mark each)

(b) The force that keeps objects such as Pluto in orbit is the gravitational force of the Sun. (Accept gravity.) (1)

(c) The information in the question suggests that Pluto was discovered so late because it is very small (1) and very far away from Earth (1), which made it difficult to spot. Telescopes were not good enough to allow astronomers to see Pluto earlier. (1) (Any two good reasons required.) (2)

9 (a) A shadow is made when light is blocked (1) by an opaque object. (1) (2)

(b) Moving the puppet away from the screen will make the shadow bigger (1) and less sharp. (1) (2)

(c) Materials that allow some light to pass through are described as translucent. (1)

10 (a)

4 marks for points correctly plotted on graph grid. Deduct 1 mark for each error. (4)

(b) 1 mark for smooth curve connecting the points. (1)

(c) The mass of the cloth after 15 minutes is 23 g (or as shown by the line drawn on the graph).
(1 mark for correct answer as shown on the graph, 1 mark for showing construction lines on graph
to show how the answer is found) (2)
(d) The mass of the dry cloth is 12 g. (You can tell this because the mass does not drop below this figure,
showing that no more water is being lost.) (1)
(e) The cloth in the shady place would dry more slowly (1) because it would be cooler (1) and therefore
the water would evaporate more slowly. (1) (3)

Science Practice Papers published by Galore Park